Questioning the Carrier

Questioning the Carrier

Opportunities in Fleet Design for the United States Navy

JEFF VANDENENGEL

Naval Institute Press
Annapolis, Maryland

Naval Institute Press
291 Wood Road
Annapolis, MD 21402

Library of Congress Cataloging-in-Publication Data
Names: Vandenengel, Jeff, author.
Title: Questioning the carrier : opportunities in fleet design for the U.S. Navy / Jeff Vandenengel.
Other titles: Opportunities in fleet design for the U.S. Navy
Description: Annapolis, Maryland : Naval Institute Press, [2023] | Includes bibliographical references and index.
Identifiers: LCCN 2023002427 | ISBN 9781682478707 (hardcover)
Subjects: LCSH: United States. Navy—Operational readiness. | Aircraft carriers—United StatesHistory. | Sea-power—United States—History—21st century. | United States. Navy—Forecasting. | Navies—Forecasting. | United States—History, Naval—20th century. | United States—History, Naval—21st century.
Classification: LCC VA50 .V36 2023 | DDC 359.030973—dc23/eng/20230315
LC record available at https://lccn.loc.gov/2023002427

♾ Print editions meet the requirements of ANSI/NISO z39.48-1992 (Permanence of Paper).
Printed in the United States of America.

31 30 29 28 27 26 25 24 23   9 8 7 6 5 4 3 2 1
First printing

The views expressed in this publication are those of the author and do not necessarily reflect the official policy or position of the Department of Defense or the U.S. government.

For Briana

Contents

Figures

INTRODUCTION

It follows then as certain as that night succeeds the day, that without a decisive naval force we can do nothing definitive, and with it, everything honorable and glorious.[1]

—President George Washington

The U.S. Navy has the most powerful fleet in all naval history. The large nuclear-powered aircraft carrier is the most powerful surface warship in that fleet and in all naval history. The U.S. Navy should stop building them immediately.

Navies' selection of their force structure—the type and number of ships comprising the fleet—is of critical importance, and that is especially true for the U.S. Navy today. A strong fleet both deters adversaries from waging wars, and wins those wars when deterrence fails. However, as renowned naval strategist Capt. Alfred Thayer Mahan wrote, a strong fleet does so much more than search for "the sterile glory of fighting battles merely to win them."[2] Mahan showed that a strong navy protects friendly seagoing commerce, which provides the wealth and stability needed for the nation to increase its power and advance its goals. Thus, the debate over the U.S. Navy's force structure cannot be restricted to ships' wardrooms and the Pentagon. The fleet's success is of national importance, affecting everything from the price of this book to the political party in power.

Fleet design is of growing importance today because of the rise of the Chinese and Russian militaries. Gone are the days of an obsolete coastal Chinese fleet and a Russian force rusting in port. Today, the People's Liberation Army Navy has the largest surface and submarine fleets in the world and the Russian Federation Navy is fielding new technologically advanced platforms and weapons on par with those of the United States.[3] If the U.S. Navy cannot retain its dominant status, the Chinese may come to control the western Pacific and its vital commerce, with direct effects on relative Chinese and American wealth and power. Similarly in the Atlantic, an ineffective U.S. Navy would reduce the strength of the North Atlantic Treaty Organization (NATO), causing shifts in European trade patterns, again diminishing the nation's wealth and power. If the U.S. Navy designs its fleet poorly going forward, its weakened ability to deter or win wars will affect the entire nation, just as the degradation of the Royal Navy in the twentieth century led to Britain's decline in influence and status.

Throughout naval history the task of designing a fleet to protect commerce and win wars has often proved difficult. Ships' long operating lives means navies must build platforms and weapons that remain viable over a span of decades. This is a challenging task under normal conditions, but rapid technological advances—such as those represented by USS *Monitor* and HMS *Dreadnought*—sometimes render entire fleets obsolete virtually overnight. Yet when trying to overcome that challenge and predict how naval warfare will evolve over a ship's lifespan, there are few robust tools guiding fleet planners. Wargames, threat analyses, and expert opinions can provide clues, but they are often inaccurate predictors of the future of naval warfare. Recent combat is a better guide for what does and does not work at sea, but even that is backward looking and tends to be misleading and open to interpretation.

The challenge of designing a fleet is even more difficult today. Technology is advancing faster than ever before, spurred on by factors like Moore's Law and global communication networks. As designers incorporate that new technology into warships, the ships become more complex and take longer to build. Thus, while technology advances faster, ship construction slows down, widening the gap between the two and making it difficult for even new ships to remain modern for long. More seriously, with essentially no conventional naval combat since World War II, it is very difficult for navies to know what platforms and weapons will prove successful in future fights at sea and which are already obsolete.

For decades, the U.S. Navy's answer to the question of fleet design focused on the large nuclear-powered aircraft carrier (CVN). Its awesome firepower, incredible endurance and logistics capabilities, and unmatched flexibility enabled it to accomplish a wide range of peacetime and wartime missions. By deterring numerous adversaries across the globe and deploying its powerful air wing when that deterrence failed, the carrier protected the nation's trade and enabled the country to thrive. The aircraft carrier is not only a key reason for the United States' enduring status as a superpower, it is also a key symbol of that status to friend and foe alike.

Throughout that illustrious history the carrier always had flaws, including its expense, vulnerability, short-ranged air wing, and its centralization of a great deal of power onto a single ship. The Navy continued to rely on the carrier because its many advantages outweighed those flaws and there was no better option. That is no longer true. Today the U.S. Navy has the opportunity to embrace the Age of the Missile, network its distributed fleet, and diversify its kill chains like never before.[4] Those opportunities mean it is now possible to design a technologically and tactically realistic fleet that outperforms today's carrier-centric model, all within the confines of the same long-term shipbuilding budget. The Navy does not need to be bound to a platform that, while greatly improved since its inception, still operates in the same fundamental manner with the same fundamental flaws it started with more than a century ago. The carrier, for decades the reason that the U.S. Navy was so great, is now holding it back from an even better future.

When the Navy does settle on a better fleet structure that reduces its reliance on the CVN, it will not be a one-for-one replacement of ships; there is no single ship afloat that can match the large aircraft carrier in capability. Similarly, there is no single class of ship that can replace the *Nimitz* and *Ford* classes of carriers; simply substituting CVNs with increased numbers of light carriers (CVLs) will result in a weaker fleet. To move beyond the carrier, the

U.S. Navy will need to revamp how it builds and operates the entire fleet. The carrier is so central to everything the U.S. Navy does that it cannot be replaced without altering the roles and responsibilities of almost every other platform. In this evolution, what matters are not the capabilities of a single ship or class, but those of the entire fleet.[5]

Part I of this book outlines what changes the U.S. Navy should make to capitalize on new opportunities in fleet design. It outlines a hypothetical fleet to show that it is financially, technologically, and operationally realistic to develop a force that is more capable than the carrier-centric model. Part II examines why the nation should make those changes to the fleet structure now. If the carrier-centric force is going to remain a viable option in the future, the Navy would be best served by continuing to use that force structure to avoid the cost and time required to transform the fleet. However, a review of naval history and the present status of the world's navies shows that there are too many ways the nation's adversaries can deter, damage, or destroy a carrier—and thus defeat the U.S. Navy. Finally, Part III examines how the Navy can evolve its fleet structure, as well as some of the difficulties it will likely encounter when it does. The decision to transform the fleet's structure and cancel billion-dollar contracts will likely incur fierce resistance from some military officers, defense contractors, and political leaders. Overcoming that resistance and the many challenges associated with evolving the fleet structure will require Navy leadership that is, as a former vice chairman of the Joint Chiefs of Staff said, "noisy, impatient, creative, courageous, and insistent."[6]

As an active-duty naval officer, I know that publishing this book may not seem to be a great career move. A book with a provocative title that examines contentious topics like force structure, fleet defenses, and defense contractors' employment of retired flag officers would seem to be a surefire way to earn a one-way ticket to some remote island. However, I doubt that is the case—and that is a credit to the Navy. The U.S. Navy has a long history of fostering constructive debate and overcoming inertia to improve itself. I wrote this book not to embarrass or impugn, but to improve the Navy that I love, a great force that I want to become even better. I am confident that the combination of a genuine motivation with a strict compliance to the Department of Defense's (DOD) publishing guidelines means this work will have no effect on my career. The Navy's leaders may not agree with my views, but they have historically been supportive of those willing to constructively offer their recommendations.[7] I expect the same response to my work.

Many better officers and leaders than this writer have already contributed to the carrier debate, including noted admirals, such as Bradley Fiske, Chester Nimitz, and Michael Gilday, as well as several presidents, included Theodore Roosevelt, Harry Truman, and Joseph Biden. This book will not end that debate, but it does seek to be the beginning of the end—of both that debate and of the supercarrier. Regarding other similar naval policy debates, Mahan wrote, "The study of the Art and History of War is preeminently necessary to men of the profession, but there are reasons which commend it also, suitably presented, to all citizens of our country. Questions connected with war . . . are questions of national moment, in which each voter—nay, each talker—has an influence for intelligent and adequate action, by the formation of sound public opinion; and public opinion, in operation, constitutes national policy."[8]

Of all the "questions connected with war" facing the U.S. Navy today, none is more important than the carrier question. Whether you are an admiral, sailor, senator, or citizen, know that your opinion has an impact on the carrier debate—and thus our nation's commerce, wealth, and power. As you read this book and develop your own opinions about the U.S. Navy's opportunities and risks in fleet design, I think you, too, will come to question the carrier.

PART I

OPPORTUNITIES

What determines the obsolescence of a weapon is not the fact that it can be destroyed, but that it can be replaced by another weapon that performs its functions more effectively.[1]

—Fleet Adm. Chester W. Nimitz, USN

CHAPTER 1

OPPORTUNITIES IN FLEET DESIGN

The tactical maxim of all naval battles is attack effectively first.[1]

—Capt. Wayne P. Hughes, USN (Ret.)

The age of the aircraft carrier is ending, not because of the ship's flaws but because of its opportunity cost: a restructured fleet better able to accomplish the U.S. Navy's missions. The carrier's flaws are growing, but they are not new. Carriers have always been expensive and somewhat vulnerable. They have always featured an air wing with short ranges and centralized a great deal of power in a single ship. For more than seventy years, the U.S. Navy accepted those flaws because the ship's many benefits outweighed the risks and because there was simply no better alternative. That is no longer true.

The advent of missiles means that nearly every ship in a fleet can use the era's premier weapon, instead of having them limited to capital ships. Modern networks enable fleets to accurately share more information faster and over longer distances than at any point in naval history. Fleets can diversify their kill chains to unprecedented degrees, incorporating platforms under the sea, on the surface, in the air, in space, and in cyberspace. Advances in processing power, reconnaissance, and communications, as well as technologies decoupling ship size and capability, enable these opportunities. They allow for a force structure that performs the carrier-centric fleet's functions more effectively using a navy consisting of more platforms with less total risk and within the same budget.[2]

These fleet design opportunities were identified years ago. The U.S. Navy did not act on them at the time because with essentially no credible adversaries at sea, it determined that it could continue to rule the seas with carrier strike groups and did not need to incur the expense of changing the fleet's structure. However, that situation has changed. With the rise of the Chinese and Russian militaries, the U.S. Navy can no longer afford to be simply good. It cannot maintain its dominance without seizing on modern opportunities in fleet design, each of which offers fundamental improvements over today's force structure. Even in cases where the Navy has made progress on taking advantage of these opportunities, it has been unable to fully attain their maximum potential while maintaining the carrier-centric model; in each case, they are mutually exclusive.

Seventy years ago a similar debate raged on the utility of the battleship; that vessel suffered from many of the same flaws and vulnerabilities as the carrier. However, as naval architect Dr. David Brown wrote, "It is often said that the battleship died because it was vulnerable. This is incorrect; it was replaced by the fleet carrier which was much more vulnerable. The battleship died because it was far less capable than the carrier of inflicting damage on the enemy."[3]

The carrier is now dying because it is less capable of "inflicting damage on the enemy" than new fleet structures. This chapter examines opportunities for a better fleet and how the U.S. Navy's carrier-centric model is prevented from capitalizing on them. Chapter 2 then proposes an example of a better fleet structure designed to seize on those opportunities. Concluding Part I, Chapter 3 evaluates that new fleet's ability to execute the U.S. Navy's missions. It shows that the carrier's waning utility and not its growing vulnerabilities is the root of its problems.

Following World War II, critics argued that the carrier was obsolete because nuclear weapons could destroy them. Admiral Nimitz countered their argument saying, "Vulnerability of surface craft to atomic bombing does not necessarily mean that they have become obsolete. What determines the obsolescence of a weapon is not the fact that it can be destroyed, but that it can be replaced by another weapon that performs its functions more effectively."[4] At the time, the aircraft carrier was not obsolete because despite its vulnerabilities, there was no better way to fight at sea. It was the most powerful and versatile vessel afloat and the best choice to lead the U.S. Navy. Since then, the carrier's vulnerabilities have remained while new options in fleet design have arisen. The aircraft carrier is growing obsolete not because of those vulnerabilities, but because it can now be replaced by an alternative fleet structure that "performs its functions more effectively."

RETURN TO "ATTACK EFFECTIVELY FIRST"

In land warfare, the relative advantage has shifted between the attacker and defender throughout history. For example, it was generally favorable to fight on the defensive in the Civil War and World War I, but improved weapons and mobility in World War II shifted the advantage to the attacker. Despite these shifts on land, at sea the attacker has held an inherent advantage throughout naval history. The attacker can concentrate its force while the defender may be spread out, can choose the timing of the battle, and can inflict decisive damage before the enemy ever gets to fight back. This idea is the theme of Capt. Wayne Hughes' authoritative book, *Fleet Tactics and Coastal Combat*, which asserts, "The tactical maxim of all naval battles is Attack effectively first."[5]

The Battle of Midway is an excellent illustration of Captain Hughes' rule. Armed with the knowledge of Japanese plans from American codebreakers and better scouting, Adm. Frank Jack Fletcher ordered the first carrier strike against the Japanese fleet. Three ships—USS *Enterprise*, USS *Hornet*, and USS *Yorktown*—launched dive-bombers, torpedo-bombers, and fighters that sank the Japanese carriers *Akagi*, *Kaga*, and *Soryu*. Although the battle continued after that first strike, with the Japanese able to sink *Yorktown*, the outcome was decided; the U.S. Navy had won its greatest victory by attacking effectively first. The first attack was not even particularly impressive; the U.S. Navy benefited from several lucky breaks and only 7 percent of its dive-bombers and torpedo-bombers hit their targets. Despite this lackluster performance, it prevailed against its formidable foe, emphasizing the benefits of the offensive in naval combat.

Over the last several decades, the U.S. Navy could afford to ignore the importance of attacking effectively first because there were simply no credible enemies to attack. With the demise of the Soviet Union, there were no powerful adversaries at sea and so the U.S. Navy appropriately shifted its focus to power projection ashore. Adversaries like Iraq, North Korea, Serbia, and Afghanistan could not challenge the U.S. Navy, which sailed the seas with impunity and reduced its investments in high-end fleet warfare.

The Navy no longer has that luxury. Facing the threat of the powerful Chinese and Russian militaries, the U.S. Navy needs to shift its posture from one of defending high-value power projection platforms to one in which every ship provides sensors and weapons that helps the entire fleet attack effectively first. The geopolitical and military situation has shifted resulting in a world in which the U.S. Navy can no longer assume it will always enjoy sea control.

A carrier-centric fleet is limited in its ability to attack effectively first. Although the U.S. Navy is making strides in improving its offensive capabilities, a fleet centered around the aircraft carrier can only take that effort so far. To protect such an asset and liability, the U.S. Navy has to devote huge amounts of financial and operational resources to its defense, focusing on a myriad of "anti" mission areas: antiair, anti-surface, antisubmarine, and anti-mine defense. All of those platforms and resources focus on defending the carrier; they are not focused on attacking effectively first, and as Captain Hughes wrote, "Defense does not dominate battle at sea and has seldom been more than a temporizing force."[6] As the U.S. Navy realigns itself for Great Power competition, it should strive toward regaining the ability to always attack effectively first. It cannot do that while centered around the aircraft carrier.

OPPORTUNITY #1: EMBRACE THE AGE OF THE MISSILE

Throughout naval history, there has been a close relationship between the era's primary weapon and the capital ship that carried it—simply due to physics. Only massive multi-deck ships like HMS *Victory* could deliver a devastating broadside. Only battleships like USS *Missouri* could carry large-caliber guns, such as her 16-inch main battery. Only aircraft carriers such as USS *Gerald R. Ford* (CVN 78) can launch advanced aircraft like the F/A-18 and F-35.[7]

As a result of this close relationship, only a small portion of a navy's ships have historically been able to use the fleet's primary weapon. At the end of World War I, only 19 percent of the U.S. Navy's surface warships could employ the era's primary weapon, the large-caliber gun on battleships. At the end of World War II, 13 percent of the U.S. Navy's major surface combatants could use one of the era's primary weapons, either large-caliber guns or aircraft.[8] During Operation Desert Storm, 8 percent of the U.S. Navy's major surface combatants could use the fleet's primary weapon, the aircraft.[9]

The rest of the fleet, unable to use the most powerful weapon of the day, still played an important role. They supported and defended capital ships, executed their own kill chains, and performed a variety of auxiliary roles. In the Age of Sail, frigates scouted for the enemy and relayed messages but did not typically enter battle with the ships of the line. In the early twentieth century, destroyers and cruisers protected battleships from torpedo and submarine attacks and searched for the enemy, but did not typically engage enemy battleships themselves. Guided-missile destroyers and cruisers focus on defending aircraft carriers from enemy attacks, enabling them to carry out offensive operations. Throughout history, each supporting type of ship had its own ability to attack the enemy, but they were generally never as powerful as that of the capital ship.

The missile changes that relationship and presents an incredible opportunity for the U.S. Navy. Instead of fleets in which fewer than 20 percent of the warships are armed with the era's primary weapon, the U.S. Navy could install missiles on 100 percent of its surface warships. Even better, those missiles could also be mounted on submarines, amphibious vessels, support vessels, aircraft, and even trucks. Those smaller ships could still perform their historical role of supporting capital ships and executing auxiliary functions, but they could also launch their own attacks on the enemy with the exact same weapon system as the fleet's most capable ship. It is a rare opportunity in history and one that would enable a fleet to better attack effectively first.

During World War II, it would have been suicidal for a small ship like an antisubmarine patrol craft to attack a battleship. The disparity in sensors, armament, and most importantly weaponry was so huge that the battleship would be virtually guaranteed a victory. Now a duel between a patrol craft and guided-missile cruiser would be a closer matchup. The cruiser, with better sensors and defensive systems, would still be more likely to prevail. However, because both ships can use the same weapon, the cruiser's likelihood of victory is much smaller than that of the battleship seventy years ago. The missile is the great equalizer.

Missiles are attractive not just because of their platform flexibility, but also because of the advantages they confer to the attacker. They require minimal investment and have a large payoff. Their operation is technically simple, with even irregular forces like the Houthi rebels being able to employ them.[10] They do not require local air or maritime superiority to employ and they can be launched with little tactical, human, or financial risk—unlike a manned aircraft. With adequate targeting information they are highly accurate, as demonstrated by a single Argentinian fighter that launched an Exocet missile and sank a British destroyer in the Falklands Conflict.[11] Most importantly, attackers can employ missiles in large quantities, which enables saturation attacks, maximizes the chances of target destruction, and allows for attacks even when the chances of success are low.

While missiles are cheap and easy to launch, they are expensive and difficult for a defender to defeat. Defenders must be prepared for attacks from a variety of platforms coming from any sector with little if any notice.[12] When the frigate USS *Stark* was tracking an Iraqi F-1 Mirage in the Persian Gulf in 1987, none of the ship's radars detected the incoming Exocet missile and the ship did not know it had been fired upon until a lookout spotted the incoming missile seconds before impact.[13] Once detected, destroying an incoming missile that may be supersonic, is flying at sea-skimming altitudes, has a very small radar cross section, and is capable of evasive maneuvers is like shooting a bullet with a bullet.[14] That task is made even harder because defenders may not know the type of incoming missiles, and radar jammers have no impact on infrared seekers. As a result, navies must rely on increasingly expensive soft- and hard-kill missile defense systems that must operate flawlessly to serve their purpose.

Even if the defender can detect, track, and destroy the incoming threat, they can still lose the quantity game. For example, if an interceptor defense system is 90 percent effective, a figure much more optimistic than what has historically been achieved, then a salvo of just ten missiles has a 65 percent of hitting the target once.[15] Once a hit does occur, history has shown that missiles' high speeds and advanced munitions often result in the ship suffering a mission kill or sinking. Capt. Arnold Henderson understood the menace of the anti-ship cruise missile (ASCM), writing, "As surface warriors will tell you, the threat that keeps them awake nights is the Supersonic Sea-Skimming Anti-Ship Cruise Missile."[16]

Missiles are far from perfect weapons. They require accurate over-the-horizon targeting, typically force ships to return to port to rearm because of the difficulty of reloading missiles at sea, and do not lend themselves to effective battle damage assessment. These obstacles are part of the reason that the U.S. Navy "has until recently all but ignored the potential of the antiship missile," according to Capt. Robert Rubel, a retired naval aviator and Naval War College professor.[17] For years the U.S. Navy dominated the world's oceans using its carrier strike groups, so there seemed little reason to undergo the effort and expense of transforming the fleet's structure to embrace missiles.

That logic no longer holds true. The Russian and Chinese navies have made prolific use of missiles. Their capabilities, quantity, and varied launch platforms threaten the U.S. Navy's dominance and challenge the assumption that the carrier-centric fleet is the best way to fight at sea. As Cdr. Phillip Pournelle, an alumnus of the Pentagon's revered Office of Net Assessment, wrote, "The age of uncontested seas is coming to an end, and ASCMs are sounding its death-knell."[18]

The Navy has enjoyed incremental success in better capitalizing on missiles' potential, pushing to upgrade designs and install them on more ships.[19] While these are a step in the right direction, they are far from a full embrace of missiles as the fleet's primary weapon.[20] The U.S. Navy cannot fully capitalize on the opportunities of the Missile Age while remaining centered on the carrier, a ship not focused on that weapon. To best take advantage of the opportunity presented by missiles, the U.S. Navy needs to shift away from a carrier-centric fleet toward a structure better able to use the era's premier weapon. No competent army would have fought in the Bronze Age without relying on bronze weapons, and today no competent navy would fight in the Missile Age without relying on missiles.

OPPORTUNITY #2: NETWORK THE DISTRIBUTED FLEET

Throughout naval history, a force's superior ability to scout the enemy and order an attack has often been crucial. Captain Hughes wrote, "At sea better scouting—more than maneuver, as much as weapon range, and oftentimes as much as anything else—has determined who would attack not merely effectively, but who would attack decisively first."[21] After spending weeks searching for the French fleet before finding and defeating it at the Battle of the Nile, Adm. Horatio Nelson is alleged to have written, "Was I to die this moment, want of frigates would be found stamped on my heart!"[22] More than a century later, while discussing mission command and the command and control that can decide a battle once the enemy is found, Adm. Arleigh Burke said, "The difference between a good officer and a poor one is about ten seconds."[23]

Much of that struggle to find and strike the enemy boils down to a rough comparison of sensor and weapons ranges. In the Age of Sail, fleets could typically see the enemy well before they could fire at it; sensor range was much greater than weapons range. A small increase in weapons range could be pivotal, such as in the Battle of Valparaíso (March 28, 1814) when two British warships used their 18-pound cannon to batter USS *Essex* from outside the range of her carronades.[24] Later, in the era of the battleship, surface warships' sensor and weapons ranges were roughly equal. Any incremental increase in the range of either could be decisive, leading to the use of tools like early fire control systems, wireless communications, and spotter aircraft. With the introduction of the aircraft carrier, sensor and weapons range expanded dramatically and it became even more important to find the enemy first. For example, at the Battle of Midway, superior American scouting combined with inferior Japanese command and control—ordering and re-ordering different aircraft ordnance types—enabled the U.S. Navy to attack effectively first and win the battle. As naval theorist and Italian admiral Guiseppe Fioravanzo wrote, "The fundamental tactical position is no longer defined by the geometric relationship of the opposing formations but by an operational element: the early detection of the enemy."[25]

As weapon ranges continue to increase, early detection becomes more important yet more challenging. For every doubling in weapon range, the targeting area that must be scouted quadruples.[26] No matter the range, speed, or lethality of a missile, there is little reason to fire it without targeting data, and that depends on how fast and accurately fleets can track the enemy and order an attack. In many ways, weapons range now exceeds sensor range.

Fortunately, modern sensor and communication technologies present an opportunity for the U.S. Navy to track the enemy at unprecedented ranges, speeds, and accuracies, thus increasing its sensor range to take advantage of today's long weapons ranges. With existing technology, the U.S. Navy can field an interoperable, fleet-wide network that seamlessly shares targeting data across all the Navy's platforms and headquarters. At the tactical level, instead of each ship needing to find, track, and develop a firing solution for the enemy, only one ship needs to do so, accelerating the rate at which the fleet can attack while reducing the risk of being found first. At the operational and strategic level, commanders would have accurate common operating pictures that reduce confusion and better inform their decisions. That network would be better prepared to integrate with unmanned and autonomous vehicles as they become more widespread and robust. Most importantly, a universal network enables a distributed fleet and diversified kill chain. In that structure the overall network matters more than individual platforms.[27]

When Cdr. Frederick Moosbrugger led his six U.S. Navy destroyers against the Japanese in the Battle of Vella Gulf in 1943, his radar enabled him to find the enemy destroyers early and thus attack effectively first. Although he had to manually radio that radar data between ships to cue them, and each destroyer had to then independently find and target the Japanese, his scouting and command and control network was still faster than that of his adversary. That speed enabled him to sink three Japanese destroyers without losing any of his own.[28] If a similar battle occurred today, a single ship tracking the enemy could instantly share its precise targeting data with accompanying ships, missile shooters hundreds of miles away, and even the Pacific Fleet commander sitting in his office in Hawaii. That capability speeds up the fleet's ability to attack, allows more units to join in the attack, and gives commanders unmatched understanding and control of that attack.

In the decades since World War II and fights like Commander Moosbrugger's, the Navy partially capitalized on the rapid advancement in communications technology. It made huge leaps in satellite communications and developed data link systems such as Link 16 and the Cooperative Engagement Capability (CEC).[29] However, because the U.S. Navy was so much faster and more capable than its weak or nonexistent adversaries, it did not matter that those tactical data link systems were not interoperable, not available on all fleet platforms, and did not strive for universal coverage. The Navy could afford to sacrifice sensor range to save budget resources because there were no credible enemies at sea. No adversary was going to launch an attack on the U.S. Navy from even a few miles away, so there was no reason for the Navy to spend money and time tracking targets hundreds of miles away.

With the rise of the advanced Russian and Chinese navies, U.S. scouting and communications can no longer be just "good." The explosion in types and numbers of sensors in all domains means it will be increasingly difficult to keep any platform hidden for extended periods of time. If the Japanese could find and track carriers during World War II, then the Chinese can do it today, especially with the help of satellites, advanced radars and sonars, unmanned platforms, and cyberattacks. Capt. Arthur Barber III, a retired surface warfare officer and former chief analyst of future force structure for the U.S. Navy, described today's situation by writing, "The ability to detect warships at long ranges or even globally is no longer a U.S. monopoly. . . . Sensor capability is advancing faster than the ability to elude detection. Long-range precision-guided weapons are proliferating and can be brought to bear in numbers against what these sensor systems detect."[30]

With navies able to find the enemy faster than ever, and missiles enabling rapid, long-range attacks, there is pressure at both ends of the kill chain to convey tracking data and attack orders more efficiently. The Navy has already started work building more robust, all-encompassing networks, an effort led most recently by the Project Overmatch initiative.[31] However, even if the U.S. Navy can build a capable universal network to meet that demand, it cannot fully capitalize on its capabilities while relying so heavily on the aircraft carrier. With so much money and power invested in a single platform, there are simply not enough other sensors and shooters feeding and supporting such a system. It is the network, not the individual platforms, that matters. The U.S. Navy is diligently working to improve its scouting and communication systems, but a network connecting a small number of nodes will never be as powerful as one linking a large distributed fleet.

In the Age of Sail, fleets used their mobility to gain the weather gauge, an upwind position that gave them the option of disengaging or attacking. In today's world, the fleet that can find and target the enemy first will have the "scouting gauge." Their faster scouting and information network will give them the option to withdraw or attack effectively first. There are better options than the carrier-centric fleet to ensure the United States retains the scouting gauge.

OPPORTUNITY #3: DIVERSIFY THE KILL CHAIN

Throughout much of naval history, navies have relied on capital ships to execute their fleet's kill chains because they were the only ships large enough to carry each era's predominant weapons. In the Age of Sail, only ships of the line could physically carry large numbers of guns, and with their short firing range, they could more effectively concentrate firepower than an equal number of guns spread out over multiple ships. Later, only battleships had the size to carry the large-caliber guns and defensive armor needed to prevail in battle. With the advent of naval aviation, aircraft carriers again needed to be large so they could launch and recover aircraft laden with fuel and munitions. In each of these cases, fleets employed other ships to execute secondary kill chains, such as sailing frigates and submarines, but the capital ships were the primary means of destroying the enemy. However, in the Age of the Missile, physics no longer requires massive capital ships to execute fleets' primary kill chains. Unlike much of naval history, varied platforms such as submarines, small patrol craft, large surface combatants, aircraft, and ground-based launchers can all fire the exact same type of weapon as the capital ship, allowing unprecedented kill chain diversification.

Modern communications allow for even greater kill chain diversification. In the Age of Sail, ships typically sailed in a line for easy fleet control and to better concentrate their firepower. Battleships, with radio communications and better firing ranges, could sail in loose formations and still coordinate their fires onto a single target. Later, carrier groups could sail independently and then coordinate their attacks in time and space. Modern communications and missiles' long ranges continue that trend, allowing commanders to execute attacks from distant platforms on single or multiple targets at exceptional ranges and speeds. Furthermore, those platforms can coordinate to complete parts of each other's kill chain. For example, a submarine can relay targeting data to a surface ship that launches a weapon that receives mid-course updates from an aircraft. The shooter may never even detect the enemy. This cooperation between units is not new to naval warfare, as even battleships used spotter aircraft to guide their shots, but the speed, interconnectedness, and flexibility of those networks is.

This presents an incredible opportunity for the U.S. Navy. It can design a fleet built around both small and large platforms that all have the ability close the kill chain, independently or cooperatively, yielding benefits at each step in that chain. When searching for and tracking the enemy, a fleet of more platforms better spread out has improved geographic and spectrum coverage of the sea and air. Being closer to the enemy, it can better classify targets than a single sensor searching huge swaths of the ocean, such as an E-2 Hawkeye aircraft or a SPY radar. When deciding if and how to attack, a diversified fleet can better feed recommendations to central hubs, such as on capital ships, but it can also decide to attack on its own if the opportunity arises or if communications are lost. That fleet has more platforms (submarines, small and large surface ships, aircraft, unmanned platforms) launching more weapons from more domains (undersea, surface, air, space, cyberspace) from more vectors. This complicates the enemy's ability to defend against attacks and raises the chance the fleet can complete at least one kill chain to destroy the target. Finally, for those same reasons, a fleet of distributed platforms can more effectively conduct battle damage assessment and execute additional attacks if needed.

A distributed and diversified fleet is not only able to better attack effectively first, but it is also more survivable. Ships like frigates and patrol craft have smaller visual, electromagnetic, and acoustic signatures than capital ships, and their increased numbers would make them more difficult to effectively track than a few large ships with distinctive signatures. When the enemy launches an attack, that step in the kill chain is also more difficult; as Adm. Arthur Cebrowski and Stuart Johnson wrote, a fleet using numerous, dispersed, low-signature platforms and weapons will "complicate planning, decision making, and operational problems" for the enemy.[32] The enemy may ironically need to fire more munitions at small ships than capital ships because of the difficulty in tracking and hitting the small targets. That logistics problem is exacerbated because the enemy must launch munitions at large numbers of small platforms instead of just at a few capital ships.[33]

Just as there are offensive and defensive wartime benefits to operating a diversified and distributed fleet, there are also peacetime advantages. A fleet composed of more ships means commanders can accomplish more presence missions. With platforms of varied sizes, commanders can better tailor ships to specific presence missions instead of being forced to use advanced combatants like *Arleigh Burke*–class guided-missile destroyers for simple tasks like drug interdiction missions. This increased and improved U.S. Navy presence will serve as a more frequent and credible deterrent for potential adversaries. If they do break the peace, a fleet of numerous small and large platforms is better able to survive a surprise attack and the initial stages of a war than one centralized in just a few capital ships.[34]

A diversified fleet mixing in smaller platforms also protects both the Navy and the nation financially. There are more shipyards capable of small warship construction, spreading the workload and reducing the risk of relying on just a few large shipyards to build capital ships. When ships are damaged or lost in battle, as they will be, the impact on the Navy's shipbuilding budget to replace them will not be catastrophic. If a ship design or its systems proves ineffective or problematic, such as the *Zumwalt*-class destroyers or USS *Gerald R. Ford*'s elevators and Electromagnetic Aircraft Launch System (EMALS), more missions are already spread out to more ships so there is less of an impact on the fleet's capability as a whole.

In a diversified and distributed fleet, the irony is that the kill chains of each of the various platforms are weaker than that of a capital ship like the aircraft carrier. However, the enemy must disrupt a much larger number of varied kill chains instead of just the carrier's, making the fleet's overall ability to destroy the enemy much stronger. As Christian Brose, the former staff director for the chairman of the Senate Armed Services Committee, wrote, the U.S. military should be "defined less by the strength and quantities of its platforms than by the efficacy, speed, flexibility, adaptability, and overall dynamism of its kill chains."[35]

The U.S. Navy has already recognized this opportunity, but has been unable to capitalize on it. In 2015 the surface fleet introduced the concept of "Distributed Lethality" (or "Distributed Maritime Operations") that focused on increasing offensive power of all surface warships and then employing them in dispersed offensive formations called "hunter-killer Surface Action Groups." U.S. Navy surface fleet leadership, including three admirals—Thomas Rowden, Peter Gumataotao, and Peter Fanta—wrote, "A shift to the offensive is necessary to 'spread the playing field.'" They continued by arguing that Distributed Lethality would "provide more strike options to joint-force commanders, provide another method to seize the initiative, and add battlespace complexity to an adversary's calculus."[36] More recently, the Chief of Naval Operations, Adm. Michael Gilday, issued guidance for the "Novel Force" initiative, which looks to capitalize on improved networks to generate an "integrated any-sensor/any-shooter kill chain enabling long-range fires."[37]

The Navy's efforts are a great start, but there are limits to the fleet's possible diversification when "all fleet operations are variations on the theme of carriers delivering ordnance," as Commander Pournelle wrote.[38] The need to fund, operate, and defend eleven aircraft carriers means the U.S. Navy will be unable to reach the platform quantity or distributed structure necessary to truly diversify the fleet's kill chains. The three surface fleet admirals even indirectly acknowledge that Distributed Lethality operations will always be secondary to aircraft carrier protection, writing in the first paragraph of their article, "The surface fleet will always defend the high-value and mission-essential units; that is in our core doctrine."[39] That does not sound like a fleet designed to "attack effectively first."

Just as a diversified financial stock investment strategy yields better returns with reduced risk, so it is with fleet diversification. However, the U.S. Navy cannot fully capitalize on this diversification opportunity while everything is centered around the aircraft carrier; the designs are antithetical to this initiative. The opportunity cost of devoting tens of billions of dollars and tens of thousands of personnel to building and manning just eleven ships and then assigning dozens of other ships to protect them is too great; there are not enough resources, people, or energy left to effectively build a diversified fleet. To capitalize on the opportunity for diversified kill chains, the U.S. Navy needs to shift away from the carrier-centric model.

OLD IDEAS FOR A NEW AGE

Forward thinkers such as Andrew Marshall, Wayne Hughes, Robert Work, and Arthur Cebrowski already recognized the opportunities discussed in this chapter. They identified trends in technology and warfare and sought to exploit them for the benefit of the United States. They proposed innovative theories and ship designs such as the Revolution in Military Affairs, Network-Centric Warfare, the arsenal ship, and the Street Fighter missile craft. Each of these ideas could have led to a more effective fleet structure, but for the most part the U.S. Navy chose not to seize on their potential. Instead, it continues to rely on "a mix of ship types that are simply evolutionary improvements and larger versions of designs from two or more decades ago."[40]

Part of the reason the Navy did not adopt many of these ideas was because of the timing of their introduction. Marshall, Hughes, Work, Cebrowski, and others identified these opportunities in the 1990s and early 2000s, when there was no clear threat and it was difficult to justify the expense and effort of transforming the U.S. military. The Cold War was at an end and the Russian military was a shell of its former shelf. The Chinese navy was a coastal force using obsolete equipment. The United States chose to enjoy its peace dividend as the world's lone superpower by not changing the way its military would fight; there did not appear to be anyone to fight.

Thirty years later, the world has changed. To face that new reality and growing challenge, a military composed of the same types of platforms executing the same types of kill chain as decades ago is less likely to prevail against an adversary that embraces today's technological opportunities. Vice Admiral Cebrowski and John Garstka discussed those technological advances and their implications writing, "Society has changed. The underlying economics and technologies have changed. American business has changed. We should be surprised and shocked if America's military did not."[41]

America's military—especially its navy—has the chance to develop a fundamentally improved force. Because of advances in sensors, communications, and weapons, it can be composed of a distributed mix of capital and small ships working cooperatively to jointly execute diverse kill chains. It can be profoundly better than the already impressive U.S. Navy. It can be a force designed around the core philosophy that it must attack effectively first.

The Navy is already striving to profit from these initiatives within the confines of its current fleet structure. It is expanding tactical data link systems, installing anti-ship missiles on more platforms, and developing an array of small unmanned platforms. However, as long as the U.S. Navy remains wedded to the carrier-centric fleet, these improvements will be incremental instead of fundamental. The overall means of fighting the fleet will remain the same, just marginally improved. It is antithetical to the doctrine of a diversified, distributed, and networked fleet to be centralized on the carrier strike group. The Navy cannot have a missile-centric fleet when the primary capital ship is designed around another weapon, the aircraft. The fleet cannot be designed to attack effectively first while focusing most of that fleet on the defense of the aircraft carrier. Arguing that the U.S. Navy can maintain the carrier at the center of everything it does while also seizing on these more progressive initiatives is like pretending that putting a seaplane on a battleship was the best use of aircraft at sea in the 1930s. The improvement is incremental, not fundamental.

Seventy years ago the aircraft carrier eclipsed the battleship as the premier ship at sea. That change happened not because of the battleship's vulnerabilities, but because of the carrier's opportunities. After that transition, the battleship continued to play an important role, just not as the flagship of the fleet. Today, the carrier is similarly growing obsolete not because of its flaws, but because of alternative fleet structures' opportunities. The carrier's strengths mean it will remain a key part of the U.S. Navy for decades to come, but its weaknesses mean it can no longer be the focal point of all fleet operations. To achieve fundamental improvement, the U.S. Navy must shift away from the carrier-centric model. It is holding the fleet back at a time that there are so many incredible options on how to move forward.

CHAPTER 2

THE FLEX FLEET

If we had a clean sheet of paper, what would the Navy look like today? What could it look like tomorrow?[1]

—Secretary of the Navy Kenneth Braithwaite

On November 14, 1910, exhibition pilot Eugene Ely climbed into his biplane aboard the cruiser USS *Birmingham*. His Curtiss "Hudson Flyer," fitted with floats in case it hit the water, was on a wooden ramp that sloped downward over the ship's bow. With the engine at full power, he gained speed as he rolled down the fifty-seven feet of ramp ahead of him, but at the end his plane dropped down toward the water just thirty-seven feet below. His wheels, floats, and propeller tips touched the water, yet he was able to maintain enough speed to complete his takeoff. As he wiped water off his goggles, *Birmingham* sounded her whistle in celebration. For the first time in history, an aircraft had taken off from a warship.[2]

Despite Ely's achievement, the U.S. Navy would not be ready to restructure its fleet around the aircraft carrier for another three decades. There was a big difference between being able to launch a single fragile biplane from a makeshift wooden platform and being ready to restructure the entire U.S. Navy around a fundamentally new way of fighting. It took time to develop the technologies, tactics, personnel, and infrastructure necessary for the U.S. Navy to shift its flagship from the battleship to the aircraft carrier.[3] When a mission arose that demanded the carrier's capabilities—sea control and power projection against the Japanese Empire—the U.S. Navy's thirty years of hard work, led by innovators such as Ely, meant it was ready to restructure the fleet to meet the challenge. As historian and retired Navy captain Jerry Hendrix wrote, it is the "mission as much as the march of technology that dictates the development of an effective Fleet structure."[4]

Likewise, today's technologies that enable advanced missiles, distributed fleets, and diversified kill chains have existed for decades. The fact that the Navy has not embraced these advances to fundamentally restructure its fleet is both a function of the need for time to develop those means of fighting and of not having a mission that required such a shift. Conflicts in places like Serbia, Afghanistan, and Iraq demanded a focus on power projection ashore by a fleet unchallenged at sea, making the large nuclear-powered aircraft carrier the best option. Today, with the rise of the Chinese and Russian militaries, the mission is shifting back to sea control. Fortunately, years of the U.S. Navy's work advancing network, platform, and missile technologies means it is ready to capitalize on the opportunities identified in Chapter 1 to restructure its fleet and meet this new threat.

This chapter translates those abstract opportunities into a concrete example of a new fleet structure. That proposed fleet is not intended to be a perfect blueprint for the U.S. Navy in the years to come. Instead, it is only designed to show that the aircraft carrier is keeping the U.S. Navy from a stronger and more flexible future. That proposed fleet is an imperfect example of the possibilities awaiting the U.S. Navy when it acknowledges that the aircraft carrier's utility is waning. This proposed fleet's primary purpose is to prove that it is financially, operationally, and technically possible to design a fleet structure that outperforms today's carrier-centric model.

For lack of a better name, let that proposed fleet be known as the "Flex Fleet": a fleet composed of powerful ships numerous enough to better flex its muscles on presence missions during peace, yet distributed enough to flex without breaking when it sustains the inevitable losses of war. In mechanical engineering, there tends to be an inverse relationship between metals' strength and ductility, or flexibility. Metals which are incredibly strong tend to be brittle and break shortly after a flaw develops, whereas metals with less strength generally have more flexibility and will yield slightly without undergoing failure. Today's carrier-centric navy is incredibly strong, but if a flaw develops—an aircraft carrier is somehow unable to perform its mission—then it may also prove to be overly brittle.

Engineers overcome this brittleness problem by mixing various metals to optimize the resulting alloy's strength and ductility. Similarly, the U.S. Navy can use a better mix of small and large ships to increase the fleet's flexibility while retaining the aircraft carrier's strength for when it is needed. Alloys tend to be much better building materials than pure metals, and a fleet built from both small and large platforms is better than an all-capital ship navy.

MISSIONS AND METHODOLOGY

To design a revamped fleet structure, it is important to first know what functions the aircraft carrier and its strike group currently perform. During peacetime, these ships conduct a wide array of immensely important functions that can be summed up as "presence missions." These can include exercises with foreign nations, port calls, freedom of navigation operations, humanitarian assistance and disaster relief efforts, anti-piracy and drug interdiction operations, and intelligence collection missions.[5]

The aircraft carrier, as well as the rest of the U.S. fleet, allows American presidents to "speak softly and carry a big stick."[6] One of the best examples of an aircraft carrier presence mission came during the third Taiwan Strait Crisis (1996). The People's Republic of China attempted to dissuade Taiwan from moving toward independence by conducting military exercises and by firing missiles into the sea near Taiwan. Then–president Clinton sent two carrier groups to the region in a show of force that reassured Taiwan while reminding China of the might of the U.S. Navy.[7] To improve upon the carrier-centric fleet, the alternative must be able to better perform these presence missions, especially in an era in which "preventing wars is as important as winning wars."[8]

In war, the aircraft carrier is capable of a wide array of functions. Any fleet structure that is going to replace the carrier-centric navy must be capable of performing these functions even better, which is no small task. Its primary missions can be simplified as reconnaissance, anti-surface warfare, antisubmarine warfare, integrated air and missile defense, and strike warfare.[9] These missions can be levied against a wide array of adversaries, from unorthodox antagonists such as Iran to powerful blue-water forces like China. Chapter 3 evaluates the Flex Fleet's ability to perform those missions compared to that of the current carrier-centric model, showing that the large nuclear-powered aircraft carrier is no longer the best means of accomplishing the Navy's missions.

The Flex Fleet was developed following the simple ground rules of being technically realistic, financially feasible, and only incorporating a reasonable number of new classes of ships. First, it does not rely on any technologies not currently available. It would be easy to build a better fleet using tools such as rail guns, directed energy weapons, and autonomous warships controlled using artificial intelligence and machine learning (AI/ML), but none have proven to be feasible, at least for now. If those or other technologies mature, this proposed fleet would be in a good position to incorporate them.

Second, the Flex Fleet was designed to require the same average shipbuilding and conversion (SCN) budget as what the U.S. Navy is currently proposing on spending. Again, it would be easy to build a better fleet with a substantially larger budget, but Congress is unlikely to fund that. To accomplish that budget parity, this analysis used the U.S. Navy's battle force inventory goal for each ship type for 2049, the last year in its 2020 "Long-Range Plan for Construction of Naval Vessels," also known as its thirty-year shipbuilding plan.[10] Then, using individual platform costs and service life estimates, it calculated what the average annual SCN budget would be to continuously maintain that number of ships in the fleet. For example, if the Navy aimed to have twenty cruisers in 2049 and each cost a hypothetical $3 billion with a forty-year service life, then it would require $1.5 billion in shipbuilding costs per year to sustain that inventory. Applying this analysis to both the U.S. Navy's current shipbuilding plans and that of the Flex Fleet is a simple and effective means of comparing the fleet structures' finances.

Individual ship costs and service lives for existing ships are drawn from the Congressional Budget Office's analysis of the Navy's 2020 Shipbuilding

Plan; those for new classes are from comparable examples. Exact figures and how they were determined are annotated in each ship class' profile below. While this proposed fleet is designed to require the same annual shipbuilding finances, transitioning to that new steady state of production would be temporarily expensive, as each new class would have one-time research, development, and initial build costs. All figures are in 2019 dollars.

The Flex Fleet was designed to achieve its goals using the smallest possible impact to the Navy's shipbuilding plans. Every new class of ship has large development and initial construction costs. Large numbers of new classes of ships would make it extremely unlikely that the Navy would have the financial resources or political capital to execute such a massive shift in fleet structure. For example, in 2017 a prominent defense think tank proposed an alternate fleet structure that included five new classes of ships and dramatic increases in the numbers of unmanned platforms, yielding an impressive design that was unlikely to ever be adopted.[11] In contrast, the Flex Fleet achieves it goals while only introducing two new classes of ships and modifying an existing one.

Finally, the Flex Fleet only changes the Navy's blue-water surface fleet structure. For simplicity, it does not change the platform types or building plans for amphibious warfare vessels, fast attack or strategic deterrent submarines, the brown water Navy, the composition of the carrier air wing, the unmanned fleet, or auxiliary vessels other than those needed to supply the Flex Fleet. There are opportunities for improvement in many of those areas, but they are beyond the scope of this work and not necessary to examine the future utility of the aircraft carrier.

SHIP DESIGNS AND FINANCES

To implement the Flex Fleet, the following platforms should be added, modified, or removed from the Navy's shipbuilding plans. The Navy would keep all existing ships for the entirety of their service lives and continue with its current plans for any platforms not mentioned. To keep with the analogy of mechanical engineering properties of strength and flexibility, all new classes are named after common metallic alloys (there is no connection between these designations and the actual metal used in the ships' construction).

Steel-Class Corvettes (DDCs)

One of the Flex Fleet's most important ship classes is the Steel-class corvette. It is the direct descendent of the Street Fighter concept (promoted by then–vice chief of naval operations Adm. Don Pilling) as well as the original Naval Postgraduate School Sea Lance design sponsored by then–Naval War College president Adm. Arthur Cebrowski.[12] The Flex Fleet model has a displacement of 600 tons, top speed of 35 knots, a low radar cross section, and a range of 3,000 nautical miles (nm) at 18 knots cruising speed. It is armed with 8 Harpoon (or newer type) anti-ship cruise missiles (ASCMs), as well as a 57-mm gun and active soft-kill and short-range hard-kill defensive systems.[13] It is equipped with a short-range dual-purpose surface- and air-search radar as well as a commercial BridgeMaster or Furuno radar that will have the same electronic signature as nearby trawlers' and merchants' radars. The corvettes will have some antisubmarine capabilities, but will rely on their small size, speed, and frequent maneuvers to evade submerged attacks. Foreign designs that may prove instructive include the Singaporean *Victory* class, the Swedish *Visby* class, the Taiwanese Hsun Hai class, and the Chinese Houbei class.[14]

For short- and mid-range surveillance and for targeting data, Steel-class corvettes will carry an MQ-8 Fire Scout or similar unmanned aerial vehicle (UAV). Fire Scout helicopter UAVs have a range of approximately 150 nm, a 12-hour endurance, and a payload capacity of more than 700 pounds.[15] Because these ships will be operated in pairs, they will be able to maintain a UAV on station for extended periods by alternating each ship's Fire Scout.

Steel-class corvettes will be developed and operated to truly own the littorals. They will use their numbers, small size, and minimized signatures (visual, electromagnetic, infrared, acoustic), coupled with the Flex Fleet's network capabilities, to be the eyes and ears of the fleet. They can best enter the confused littoral environment to attack the enemy with ASCMs themselves or provide targeting data for the rest of the Navy operating farther out. They are heavily armed and capable ships, but small and cheap enough that their destruction will not be debilitating; in the event they are hit, the crew would be expected to abandon it for their partner vessel.[16] They could be employed in a wide range of wartime missions, including attacks on enemy warships, scouting, screening of larger ships or amphibious forces, or blockade duty. They are meant to go in harm's way and create havoc when they get there.

Each Steel-class corvette will cost an estimated $300 million, and the Flex Fleet will employ one hundred ten of them.[17] With an expected twenty-five-year service life, each year they will require $1.32 billion, or less than 6 percent, of the Navy's SCN to maintain their numbers. They will only be homeported in locations that could provide host nation support, to avoid the cost of tender services, similar to how the littoral combat ships are maintained in Singapore. They will be homeported in places like Japan, Guam, Singapore, Spain, Britain, and Bahrain. Although they are forward deployed, they would still be capable of crossing the ocean, just as the smaller Coast Guard Sentinel-class patrol craft are able to.[18]

Constellation-Class Frigates (FFGs)

In 2020 the Navy awarded a contract to start construction of its new class of frigates. These ships, based on an existing Italian design, will displace approximately 7,400 tons and have 32 VLS cells and 16 anti-ship missile canisters. They will focus on antiair, anti-surface, antisubmarine, and electronic warfare.[19] They will be less expensive and thus less capable than ships like *Arleigh Burke*–class destroyers and *Ticonderoga*-class cruisers. For example, the frigates will likely only provide local area antiair defense instead of the wide area capabilities of the larger surface combatants equipped with SPY radars and the Aegis combat system.

The Flex Fleet will build increased numbers of *Constellation*-class frigates, which will bridge the gap between Steel-class corvettes and *Arleigh Burke*–class destroyers. In peacetime they can steam independently or with other ships to accomplish presence missions. In war, they will be capable ships ready to execute a variety of missions, with two of their most important contributions being their missiles and their antisubmarine capabilities. Antisubmarine warfare (ASW) is a team sport requiring lots of air, surface, and subsurface platforms networked over a large area searching for and destroying enemy submarines. This need for a distributed, networked fleet structure makes the Flex Fleet an ideal choice. In the Flex Fleet, there will be a large inventory of *Constellation*-class frigates ready to accomplish missions like ASW, armed with specialized tools like

AN/SQS-62 variable depth sonar, a TB-37 multi-function towed array, and a MH-60R helicopter.[20] This will free up capital ships, such as *Arleigh Burke*–class destroyers, to perform more important missions if they are needed elsewhere.

Each frigate costs $1.1 billion and has a twenty-five-year service life.[21] The U.S. Navy is seeking to have fifty small surface combatants in 2049, which will include *Constellation*-class frigates and littoral combat ships. For simplicity, assuming all small surface combatants in the future will be *Constellation*-class frigates, those shipbuilding targets will require approximately $2.2 billion in SCN funds per year.[22] The Flex Fleet will increase the small surface combatant target inventory by 62 percent, seeking to operate eighty-one small surface combatants at an annual SCN cost of $3.56 billion.

Brass-Class Missile Arsenal Ships (MAS)

The Brass-class ships are simple, inexpensive craft with limited sensors but a large missile inventory and the ability to receive launch orders from other platforms. They are an evolved descendent of the original arsenal ship devised in the late 1990s, which was too large and suffered from many of the same problems as today's aircraft carrier.[23] The Flex Fleet variant is smaller, with 150 vertical launch system (VLS) cells, roughly the same number carried by *Ohio*-class guided-missile submarines (SSGNs) performing a similar mission. It uses a simple ship design or converted merchant hulls.[24] It can remain electromagnetically silent until it receives orders to attack, when it sails within range of its weapons at its top speed of 25 knots. With weapons like the Tomahawk, that range would be approximately 900 to 1,500 nm, a figure which will increase with newer variants and missiles.[25]

Brass-class missile arsenal ships are designed to truly take advantage of the missile era. Networked to ships like Steel-class corvettes, *Constellation*-class frigates, *Arleigh Burke*–class destroyers, and *Virginia*-class submarines, they serve as large missile batteries at those ships' disposal. Like VLS ships now, their missile inventories can be tailored to the mission. Early in a conflict they can have a mixture of anti-ship missiles and defensive interceptors to help the fleet win sea control. Later, Brass-class ships can shift their loadout to carry more land-attack missiles, providing significant firepower for power projection missions.

In the Flex Fleet, ships like Steel-class corvettes, *Constellation*-class frigates, and *Virginia*-class submarines will probe the littorals and use their own weapons to attack the enemy. However, if they run out of weapons or find large enemy fleet concentrations, and if communications are intact, they can provide coordinates to the silently waiting Brass strike ships hundreds of miles away. These ships can launch large salvos of anti-ship weapons to overwhelm and destroy the enemy, benefiting from other ships' sensors to execute a distributed and diversified kill chain. In the past, long-range ASCMs have been hampered by the need for long-range targeting and updates; for example, a sonic missile fired a thousand miles away will take seventy-eight minutes to arrive at its target, during which time that enemy ship may have moved dozens of miles away. Using the fleet's distributed platforms to provide midcourse guidance will help overcome this problem and let the strike ships remain at a safe stand-off distance. Just as the artillery provides supporting fire when required by the infantry, the Brass-class strike ships will provide additional missiles for frontline units when needed.

Brass-class ships will also be valuable in a power projection role. When attacking inland targets, missiles provide better penetration than what is available from carrier-launched aircraft, especially against defended targets. It is more difficult for the enemy to know when missiles are incoming, and a missile is much more difficult to defend against than a larger aircraft. If a missile is shot down, the Flex Fleet can simply launch another without having to suffer the tactical, financial, and political implications of a lost multimillion-dollar plane and a dead or captured pilot. As renowned naval analyst Norman Polmar said, "If you want to kill somebody, you don't send a group of F/A-18s, you have a destroyer or a submarine from several hundred miles away fire Tomahawk missiles."[26] In the Flex Fleet, Brass-class strike ships will also be launching those missiles with deadly effect.

One of the primary problems of relying on missile shooters versus the aircraft carrier is the inability to reload VLS at sea. It forces ships to leave the fight and return to port after they expend their weapons, meaning the many capabilities of ships like *Arleigh Burke* destroyers are unavailable to the fleet in the meantime. Instead of attempting to devise ways of reloading VLS missiles at sea, the Flex Fleet will replace the entire VLS cell by rotating out arsenal ships. *Arleigh Burke*–class destroyers can remain at sea and in the fight while the less capable arsenal ships switch out to reload. When paired together, the arsenal ship will benefit from the destroyers' sensors and defensive capabilities and the destroyer will benefit from the arsenal ship's large inventory of weapons. Just as metals' characteristics improve when they are alloyed with other materials, the Brass and *Arleigh Burke* classes will complement and strengthen each other.

For defense, the Brass class will primarily rely on remaining undetected far from the engagement area before it launches its weapons. Brass-class missile arsenal ships will have minimal superstructure, low freeboard, and possibly be capable of ballasting down, all to minimize their radar cross section, but only if it is cost feasible.[27] They will be capable of receiving targeting information passively. They will also have short-range air defenses such as the SeaRAM close-in weapon system.

Of course, today's U.S. Navy already has an excellent arsenal-type ship, the *Ohio*-class SSGN. The SSGNs, converted from *Ohio*-class ballistic missile submarines (SSBNs), are an ideal platform for launching Tomahawks. They can silently approach the enemy's coast, launch 154 missiles without warning, and then safely disappear.[28] In the Flex Fleet, they would be utilized throughout their remaining service lives. However, the Flex Fleet will not build new SSGNs because most of their missions can be executed by *Virginia*-class submarines or Brass-class arsenal ships and due to their estimated $7.8 billion cost.[29]

The Brass strike ships are estimated to cost $400 million each. The Flex Fleet will employ twenty-eight of them, and with an expected thirty-five-year service life, they will require $320 million, or less than 2 percent, of the annual Flex Fleet SCN budget.[30]

Arleigh Burke–class Destroyers (DDGs)

Arleigh Burke–class guided-missile destroyers are one of the most capable surface combatants in the world, with an advanced sonar suite, the Aegis combat system, dozens of VLS cells, and numerous defensive systems. However, like *Ford* and *Nimitz* carriers, the unintended consequence of relying on such individually powerful and flexible ships is that the overall fleet is less powerful and flexible because of the money required to build the Navy's few capital ships. The fleet's flexibility is further reduced because when ships such as CVNs or DDGs are lost, then their many capabilities are all lost as well.[31] The National Research Council discussed this implication in 2013:

> In a cost-constrained environment, when envisioning a low-threat operating environment or scenarios where there is little risk of loss of life or ships being damaged, building fewer but more expensive weapon systems/multimission ships is economically rational. More capability can be provided at sea with fewer hulls, crew, logistics, and total lifecycle costs. When, however, there is a risk of that ship being damaged or sunk where there is a high probability of tactical surprise attack, the reverse is true—that is, cost-effectiveness becomes "too many eggs in one basket." A damaging hit on a guided missile destroyer (DDG) hull is degradation not just of the fleet's air and missile defense capacity but also of ASW; antisurface, maritime interdiction operations; support for Marines ashore; and helicopter-related-missions as well. Building more less-expensive, single-mission ships may increase fleet resilience, to absorb the impact of an unanticipated threat at sea, and provides more options for response through geographic dispersion as well as greater ship availability for quick modifications.[32]

Relying on a fleet primarily composed of highly capable, multi-mission, expensive platforms like *Ford*-class carriers and *Arleigh Burke*–class destroyers made sense in previous decades when there were few threats at sea. Today a fleet structure highly reliant on capital ships is no longer the best choice because of their increasing vulnerability to enemy threats. When the famous captain Wayne Hughes led a group of Naval Postgraduate School (NPS) professors to propose an alternate fleet structure, they too concluded that "it is a mistake to rely heavily" on expensive guided-missile destroyers and guided-missile cruisers.[33]

To mitigate these problems, the Flex Fleet aims to spread out the destroyers' capabilities to other platforms. Other ships, such as Steel-class corvettes, *Constellation*-class frigates, and Bronze-class strike ships can accomplish some of the destroyers' missions without putting such an important and expensive ship at risk. When the need arises, the destroyers are available to reinforce the other ships and bring their impressive warfighting capability to the scene. However, with a portion of their missions distributed to other ships, there will be a need for a reduced number of destroyers in the Flex Fleet.

By 2049 the U.S. Navy aims to have 108 large surface combatants, which includes *Arleigh Burke* destroyers and its replacement.[34] Averaging the cost of an *Arleigh Burke* destroyer and its follow-on ship, each large surface combatant costs approximately $2.1 billion and has a forty-year service life.[35] To maintain that inventory, it will require approximately $5.67 billion of the U.S. Navy's annual SCN budget. The Flex Fleet, distributing many of those missions to smaller platforms, will only seek to have 75 large surface combatants, requiring approximately $3.94 billion of the SCN budget.

Ford-class Aircraft Carriers (CVN)

The second and third *Ford*-class aircraft carriers, USS *John F. Kennedy* (CVN 79) and USS *Enterprise* (CVN 80), have already been procured and are scheduled to be delivered to the Navy in 2024 and 2028, respectively.[36] Under the Flex Fleet, once those ships join the fleet the rest of the *Ford* class will be canceled. The name *Enterprise* would be used for both the U.S. Navy's first and last nuclear-powered aircraft carriers.[37] The existing carriers of the *Nimitz* and *Ford* classes will be used for the remainder of their fifty-year service lives. Under this plan, the Flex Fleet will still have these powerful ships for decades while it transitions to a new structure. There would be eleven carriers in 2030, eight carriers in 2040, five carriers in 2050, and all three *Ford*-class carriers would be serving as late as 2065.[38]

Those *Nimitz* and *Ford* carriers are still a highly capable weapons when used correctly. Instead of demanding that the carrier lead most of the Navy's missions, the diversified Flex Fleet will be better able to only use the carrier in missions it excels at. As Capt. Robert Rubel wrote, "Certain roles for the carrier are already obsolete, and others are eroding. A few new roles are emerging, but these place the carrier in a new position in relation to the rest of the fleet. Whereas the carrier has been the central pivot of the fleet since World War II, the arbiter and yardstick of naval supremacy and the keystone of fleet architecture, it will gradually become a more narrowly useful role player."[39] In peacetime, the rest of the Flex Fleet can execute most presence missions, leaving the carriers ready to assist if any dire situations arise.[40] During war they can deploy once air and maritime superiority is assured, allowing the Flex Fleet to mitigate the carrier's flaws while still benefiting from its high-volume strike capacity.

The U.S. Navy plans to have ten aircraft carriers in operation in 2049, the last year of its Fiscal Year 2020 thirty-year shipbuilding plan.[41] Each *Ford*-class carrier costs $12.8 billion and has a service life of fifty years.[42] To maintain that inventory of carriers, it will require approximately $2.56 billion of the U.S. Navy's annual SCN budget.

Bronze-Class Light Aircraft Carriers

Instead of supercarriers, the new Bronze class of light aircraft carriers will provide the bulk of the Flex Fleet's airpower. These ships are modified versions of the *America*-class of amphibious assault ships, a design the RAND Corporation examined for Congress in 2017.[43] Altered to have a larger flight deck and increased fuel and ordnance capacity, they will be approximately 40,000 tons, conventionally powered, and able to carry approximately thirty F35Bs along with up to five additional aircraft for early warning and support operations.

During peacetime, the supercarriers would give up most of their presence missions to the Flex Fleet's numerous other ships, which will serve as more frequent reminders of the U.S. Navy's power and resolve. When the nation needs to send a stronger message, it can flex its muscles even more by sending a CVL, or if needed a CVN, thus giving the nation's leaders the ability to present a more calibrated response.

During wartime, both CVNs and CVLs, relieved of their obligation to lead the fleet in virtually every scenario, will instead focus on what they do best. Under the Flex Fleet structure, the aircraft carriers will no longer be primarily responsible for missions like anti-surface warfare (ASUW)— (now led by Steel corvettes, *Constellation* frigates, *Arleigh Burke* destroyers, Brass strike ships, and *Virginia*-class submarines), ASW (now led by *Constellation* frigates, *Arleigh Burke* destroyers, and *Virginia*-class submarines), strike (led by Brass strike ships, *Arleigh Burke* destroyers, and *Virginia*-class submarines), and reconnaissance (led by a much larger dispersed and networked fleet aided by satellites and long-range land-based aircraft when possible). Aircraft carriers will still play key roles in these missions, but they would no longer be primarily responsible for their execution.

Taking some weight off the carrier's shoulders will allow it to focus on and excel at what it can do better than any other platform. In the Flex Fleet, carriers will focus on missions such as fleet air defense (mitigating the F-35's range problems), ASUW, combat reconnaissance, and high-volume strikes in benign environments. This mission realignment will also allow the aircraft carrier to mitigate many of its vulnerabilities. By not having to be within range of ground targets, the carrier can stay farther out to sea and thus reduce its chance of being located. Even if it is found, operating farther from the threat will give its defending aircraft and escorts more time to detect, engage, and destroy the threat than if they were close to shore.[44] By avoiding chokepoints and the littorals, it can also reduce the chances of enemy submarines successfully finding and tracking it, and similarly avoid the threat of naval mines, most of which can only be employed in shallow waters. Their operations would be comparable to how the British used their carriers in the Falklands Conflict. There, the carriers focused on the fleet air defense role, executing a mission they could perform better than any other platform. Due to their air wing's poor record attacking defended ground targets and the threat of Argentinian air attacks, the carriers remained far out to sea and left most other missions to the rest of the British fleet.

In the opening stages of active operations, the CVLs will provide air defense for the surface fleet, and their increased numbers will complicate the enemy's calculations while also reducing the impact of the loss of one carrier to the entire fleet. Once the Flex Fleet has won control of the seas, the CVNs arriving from the United States can enter a safer environment and bring their considerable firepower to bear.

Bronze-class CVLs have several flaws that make them individually less capable than *Nimitz* or *Ford* CVNs. CVLs have a smaller air wing that does not include the Navy's premier airborne early warning platform, the E-2D Hawkeye. Its short takeoff and vertical landing (STOVL) F-35Bs can carry less fuel and ordnance than the conventional takeoff and landing (CTOL) F-35C variant on CVNs. However, these issues are mitigated by the fleet's new structure, the carrier's reduced mission set, and their increased quantity. *Ford* carriers are equipped with forty-four F-35s and approximately twenty-five support aircraft for missions like antisubmarine warfare, search and rescue, and electronic warfare.[45] The Bronze class, by shifting many of those secondary missions to other platforms, has a reduced need for a large contingent of non-fighter aircraft yet still has a robust tactical air wing of thirty F-35Bs. The inability to operate the E-2D is important. However, there are alternative airborne warning platforms, such as Boeing's proposed variant of the V-22 Osprey tilt-rotor aircraft.[46] More importantly, a distributed Flex Fleet spread out and searching for the enemy yet networked together is better suited to perform reconnaissance and command and control than today's Navy, reliant on just a few E-2Ds flying from just a few aircraft carriers. Finally, the F-35B's range and payload reductions will be mitigated in the Flex Fleet because of their shift in focus from power projection ashore to fleet air defense and surface warfare. Instead of having to fly long distances carrying heavy ground-attack weapons, they can shift to operating closer to the fleet and using lighter air-to-air weapons.

When Captain Hughes and his NPS colleagues proposed an alternative fleet structure in their 2009 "New Navy Fighting Machine" report, they addressed the differences between STOVL and CTOL aircraft. First focusing on high-end conflicts against adversaries such as China, they wrote, "How important is the CVN's CTOL (F-18 or F-35C) range advantage over a CVL's STOVL (F-35B)? The carriers are being forced far to seaward, and so the difference in strike range probably will not matter much. Land-based aircraft will probably surpass either CTOL or STOVL range and in future air-to-air combat because of their runway advantage." Shifting from a war with a peer competitor, they wrote that in small wars, "High performance fighter-attack aircraft are probably overdesigned. An F-35B should be more than adequate because it should be able to safely perform reconnaissance and deliver strikes in almost any irregular warfare environment."[47]

Individually, Bronze-class CVLs are less capable than CVNs of the *Nimitz* and *Ford* classes. However, a fleet incorporating Bronze CVLs into a distributed structure is more capable than a fleet dependent on *Nimitz* and *Ford* CVNs. The Flex Fleet will include sixteen Bronze-class light aircraft carriers, each costing $4.3 billion (as estimated by RAND).[48] With forty-year service lives, they will annually require approximately $1.72 billion of the Flex Fleet's SCN budget.

Combat Logistics Force Ships (CLFs)

Combat Logistics Force (CLF) ships—including T-AO oilers, T-AKE dry cargo ships, and AOE fast combat support ships—are designed to have the speed, endurance, and resilience to support the fleet when it is forward deployed and conducting combat operations. The Flex Fleet, composed of more platforms spread over a wider area, will require a larger inventory of CLF ships to support it. Each CLF ship costs approximately $700 million and has a service life of approximately thirty-eight years, depending on the variant.[49] While the U.S. Navy plans to have thirty-one CLF ships by 2049, the Flex Fleet increases that target by approximately 30 percent to forty CLF ships, requiring approximately $736 million of the annual SCN budget.[50]

TURNING POINTS

When Eugene Ely took off from *Birmingham* in his biplane in 1910, he helped usher in a new era of naval warfare. The U.S. Navy realized the benefits of naval aviation and worked to transform its fleet to take advantage of that potential. That work was difficult and expensive, but after thirty years it yielded a fleet ready to win World War II. Today, the U.S. Navy has new opportunities to fundamentally improve its fleet again. Those opportunities challenge the assumption that the U.S. Navy must continue to center its forces and efforts around the large nuclear-powered aircraft carrier.

More than a century after Ely's historic flight, Secretary of the Navy Kenneth Braithwaite discussed how the Navy was at another turning point in its history and force structure. He said,

> I have asked the leadership in the Navy to join with me and kind of reimagine the Navy from the ground up. If we had a clean sheet of paper, what would the Navy look like today? What could it look like tomorrow? Not only from what ships are part of the fleet structure, what aircraft are in the air or submarines under the seas, but what's

FIGURE 2.1. U.S. Navy and Flex Fleet Shipbuilding Targets and Finances

	Ship Type	USN Target Inventory in 2049 (Ships)	Flex Fleet Target Inventory in 2049 (Ships)	Unit Cost ($M)	Unit Service Life (Years)	USN SCN/Year ($M)	Flex Fleet SCN/Year ($M)
Modified in the Flex Fleet	Corvettes Steel Class (DDC)	0	110	300	25	0	1,320
	Small Surface Combatants *Freedom* Class (LCS) *Independence* Class (LCS) *Constellation* Class (FFG)	50	81	1,100	25	2,200	3,564
	Missile Arsenal Ships Brass Class (MAS)	0	28	400	35	0	320
	Large Surface Combatants *Arleigh Burke* Class (DDG) and Replacement	108	75	2,100	40	5,670	3,938
	Aircraft Carriers *Ford* Class (CVN)	10	0	12,800	50	2,560	0
	Light Aircraft Carriers Bronze Class (CVL)	0	16	4,300	40	0	1,720
	Combat Logistics Ships (CLF)	31	40	700	38	571	737
	Large Payload Submarines *Columbia* Class (SSGN)	3	0	7,800	42	557	0
Unchanged	Amphibious Warfare Ships	35	35	2,800	40	2,450	2,450
	Ballistic Missile Submarines *Columbia* Class (SSBN)	12	12	7,700	42	2,200	2,200
	Attack Submarine *Virginia* Class (SSN) and Replacement	67	67	4,350	38	7,670	7,670
	Support Ships	39	39	350	38	359	359
	No Net Change in Annual SCN Budget						

Table Notes

1. All costs in 2019 dollars.
2. Although the Flex Fleet will not build any more CVNs after USS *Enterprise* (CVN 80), existing *Nimitz*- and *Ford*-class carriers will be used for the remainder of their service lives. Because of their long service lives, there will still be five CVNs remaining in the Flex Fleet in 2049. Source: *Bradley Martin and Michael E. McMahon, Future Aircraft Carrier Options (Santa Monica, CA: RAND, 2017)*, 2.
3. USN target inventory for 2049 from the U.S. Navy's Report to Congress on the Annual Long-Range Plan for Construction of Naval Vessels for Fiscal Year 2020. Source: *Office of the Chief of Naval Operations, "Report to Congress on the Annual Long-Range Plan for Construction of Naval Vessels for Fiscal Year 2020" (Washington, DC: Department of the Navy, March 2019)*, 13.
4. Costs and service lives for existing platforms from the Congressional Budget Office's Analysis of the Navy's Fiscal Year 2020 Shipbuilding Plan. Costs and service lives for new platforms estimated using comparable ships. Cost for Large Surface Combatants estimated using the average cost of the existing *Arleigh Burke* class and the expected Future Large Surface Combatant. Cost for CLF ships estimated using the average for T-AO-205 *John Lewis*–class oilers and T-AKE(X) replenishment ship replacements. Costs for amphibious warships estimated using the average cost of LHA-6 *America*-class amphibious assault ships, LPD-17 Flight II *San Antonio*–class amphibious warfare ships, and LPD-17 amphibious warfare ship replacements. Source: *Congressional Budget Office, "An Analysis of the Navy's Fiscal Year 2020 Shipbuilding Plan" (Washington, DC: Congress of the United States, October 2019)*, 3, 20.

the laydown of personnel? And do we have the command and control systems correct? Are they reflective of a more agile and flexible force, or are they indicative of the last war we fought and a Cold War legacy? I really believe that we are at a turning point in the history of the Navy on what the future force structure will be, and we need to make sure that that meets the emerging threat.[51]

The Flex Fleet is one alternative force structure that is "reflective of a more agile and flexible force." It is a distributed force able to execute a variety of kill chains by exploiting two of today's key technologies: networks and missiles. It is designed as an alloy of high-end capital ships and less capable role players working together to yield a stronger and more resilient fleet. It is a viable construct that does not rely on fantastical technologies or exorbitant budgets. The Flex Fleet is designed to yield the most powerful Navy possible, not the most powerful ship possible.

CHAPTER 3

THE FLEX FLEET VS.
THE CARRIER-CENTRIC FLEET

The carrier's day as the supreme arbiter of naval power and the determinant of fleet architecture may be coming to a close.[1]
—Capt. Robert C. Rubel, USN (Ret.)

Part I of this book has sought to determine if the aircraft carrier is growing obsolete. In addressing this question, we have analyzed the problem and proposed alternative fleet structures that can outperform today's carrier-centric Navy. Chapter 1 identified opportunities in fleet design that the carrier-centric fleet cannot capitalize on, including embracing the missile, networking the distributed fleet, and diversifying the kill chain. Chapter 2 used those opportunities to outline the hypothetical Flex Fleet. This alternative force structure showed that it is possible to apply those opportunities to the development of a financially and technically realistic force. We will now compare the Flex Fleet to the U.S. Navy's planned carrier-centric force. We will examine their relative ability to execute peacetime presence operations and wartime missions, as well as the secondary advantages and consequences of both fleet structures. The overall intent is to evaluate the carrier using Adm. Chester Nimitz' standard: "What determines the obsolescence of a weapon is not the fact that it can be destroyed, but that it can be replaced by another weapon that performs its functions more effectively."[2]

This analysis shows that in most operations, the Flex Fleet outperforms the carrier-centric Navy. In peacetime, the Flex Fleet has 42 percent more battle force ships able to influence the nation's partners or adversaries, allowing commanders to conduct more frequent and better tailored presence missions. In wartime, factors such as the Flex Fleet's distributed organization and its 86 percent increase in missile-shooting surface ships enable it to outperform the carrier-centric model in most fights. Finally, the Flex Fleet's smaller ships provide secondary benefits, such as reducing the burden on overworked shipyards and requiring approximately 10 percent fewer sailors at sea than the Navy's plan. Overall, the Flex Fleet is a better option in peacetime and wartime, indicating that the carrier's utility is waning.

In keeping with the analysis in Chapter 2, all comparisons to the U.S. Navy refer to the force structure planned for 2049, the last year in the Navy's 2020 "Long-Range Plan for Construction of Naval Vessels," also known as its thirty-year shipbuilding plan.[3]

A MISSION-BY-MISSION COMPARISON

The aircraft carrier's primary missions are peacetime presence missions, reconnaissance, anti-surface warfare, antisubmarine warfare, integrated air and missile defense, and strike warfare.[4] Any replacement fleet must outperform the carrier-centric model across this spectrum of operations, a difficult task.

Peacetime Presence Operations

In peacetime, the Navy executes a wide array of valuable missions that can be summarized as "presence operations." These missions serve to reassure allies, deter adversaries, and protect American interests, with examples including international exercises, humanitarian assistance missions, freedom of navigation operations, foreign port calls, and drug smuggling interdictions.

In 2016 the Navy stated that it needed 355 battle force ships to complete its required tasking, which included peacetime presence missions. However, it has generally only had between 270 and 300 ships in recent years, meaning many valuable presence opportunities often go unfulfilled.[5] In 2011 then–secretary of defense Robert Gates discussed this problem, testifying, "A smaller military, no matter how superb, will be able to go fewer places and be able to do fewer things."[6] When he was Chief of Naval Operations, Adm. Jonathan Greenert discussed how difficult it was for the Navy to meet combatant commanders' presence requirements, a problem he spent "90 percent of his time on" and that often resulted in extended deployments and shortened maintenance periods.[7]

It appears essentially impossible for the Navy to reach its 355-ship goal, and thus accomplish all required presence missions, using existing ship designs without a massive increase in the shipbuilding budget. If the Navy was going to reach this ambitious goal by 2049, the Congressional Budget Office estimated that it would require an SCN budget of $28.8 billion annually, more than double the historical average of $13.8 billion. Even compared to the larger average budgets between 2015 and 2019, Congress would still have to further increase spending by more than 50 percent.[8] The U.S. Navy chose to center its fleet around highly capable—and very expensive—warships such as *Ford*-class carriers, *Arleigh Burke*–class destroyers, and *Virginia*-class submarines. As a result the fleet is limited to relatively small numbers of those platforms with no viable path to get to its stated requirement of 355 ships. As Dr. Harlan Ullman wrote about the 355-ship goal, "There are only two ways to obtain the number: spend more money or revolutionize the way the Navy plans for the future."[9] Given the unlikelihood of the former, the latter seems to be the only viable option.

The Flex Fleet is an example of how the Navy can "revolutionize" its future force structure. Composed of less expensive ships, it is more likely to achieve the battle force inventory size needed to accomplish required presence missions. For easy comparison, assume that the Navy received the SCN budget necessary to reach 355 ships by 2049. If the Flex Fleet received the same budget, it would have a battle force inventory of 503 ships, an increase of 42 percent (as shown in Chapter 2). No matter what the SCN budget actually is, the Flex Fleet will always have more ships and thus be able to accomplish more presence missions.

In addition to accomplishing more presence missions, the Flex Fleet would be able to provide ships better tailored to each specific operation. For example, *Arleigh Burke*–class destroyers, equipped with state-of-the-art sensors and weapons, are overkill for missions like drug interdiction operations. Under the Flex Fleet structure these missions could be conducted by less capable ships such as Steel-class corvettes and *Constellation*-class frigates, freeing up destroyers for more important tasks. Additionally, the Flex Fleet, with a larger variety of more ships available at any one time, could show off its increased flexibility by more easily responding to hotspots. Commanders can deploy corvettes, strike ships, frigates, and finally destroyers and aircraft carriers, thus solving the problem at the lowest level with a customized response. Today's Navy must often choose between sending nothing or quickly escalating to capital ships, leaving few options if a stronger message is needed later.

Ship numbers are not the only factor that matters in presence missions. The type of ship involved directly affects the impact it can have, and today there is no more impactful ship than the aircraft carrier. As a brace of admirals—Robert Natter and Samuel Locklear—wrote, "Nothing demonstrates support to our friends and allies better than the arrival of an aircraft carrier and its strike group."[10] To replace the carrier-centric fleet, the Flex Fleet must outperform today's Navy both in quantity and quality of presence missions. Retired naval aviator captain Jerry Hendrix examined the quality issue in 2013, when he applied a hypothetical "presence value" to various platforms based on their capabilities. For example, he assigned aircraft carriers a presence value of 1.0 on a sliding scale, with destroyers and littoral combat ships given scores of 0.2 and 0.1, respectively. He then questioned whether it was better to achieve a 1.0 score using a single aircraft carrier, or if it could be better attained for less money by buying five destroyers for sum total of 1.0 presence.[11] Expanding this idea, Figure 3.1 uses Captain Hendrix' basis to compare the "presence value" of the ships in the U.S. Navy and those of the Flex Fleet. This is an extremely rough comparison, only intended to show that the Flex Fleet's advantage is not limited to just platform quantity. Using the estimated values in Figure 3.1, the Flex Fleet has 15 percent more "presence value" than the fleet the U.S. Navy plans on building.

Warfighting credibility is a key factor affecting ships' "presence value." Early in the history of the United States, the Barbary States seized American merchant vessels for ransom and demanded tributes because it did not view the U.S. Navy as a credible force. It was not until the United States sent multiple squadrons and repeatedly fought the Barbary States that the U.S. Navy earned the credibility to be an effective deterrent. Early in the twentieth century the

FIGURE 3.1. U.S. Navy and Flex Fleet Presence Comparison

	Ship Type	USN Target Inventory in 2049 (Ships)	Flex Fleet Target Inventory in 2049 (Ships)	"Presence Value" per Ship
Modified in the Flex Fleet	Corvettes Steel Class (DDC)	0	110	0.05
Modified in the Flex Fleet	Small Surface Combatants *Freedom* Class (LCS) *Independence* Class (LCS) *Constellation* Class (FFG)	50	81	0.1
Modified in the Flex Fleet	Missile Arsenal Ships Brass Class (MAS)	0	28	0.2
Modified in the Flex Fleet	Large Surface Combatants *Arleigh Burke* Class (DDG) and Replacement	108	75	0.2
Modified in the Flex Fleet	Aircraft Carriers *Ford Class* (CVN)	10	0	1.0
Modified in the Flex Fleet	Light Aircraft Carriers *Bronze* Class (CVL)	0	16	0.6
Modified in the Flex Fleet	Combat Logistics Ships (CLF)	31	40	0.05
Modified in the Flex Fleet	Large Payload Submarines *Columbia* Class (SSGN)	3	0	0.05
Unchanged	Amphibious Warfare Ships	35	35	0.2
Unchanged	Ballistic Missile Submarines *Columbia* Class (SSBN)	12	12	0.05
Unchanged	Attack Submarines *Virginia* Class (SSN) and Replacement	67	67	0.05
Unchanged	Support Ships	39	39	0.0
	Total Battle Force Inventory	355	503	
	Flex Fleet Net Increase in Battle Force Ships			+42%
	Flex Fleet Net Increase in "Presence Value"			+15%

Great White Fleet executed impactful worldwide presence missions because it was a credible combat force. The ballistic missile submarine force in the Cold War was an effective deterrent to nuclear attack because the Soviet Union respected it as a credible, survivable nuclear launch platform. In 1995 and 1996, carrier strike groups were effective deterrents to Chinese threats against Taiwan because they were a credible combat force.

However, after twenty-five years of developing a powerful, modern navy, the Chinese likely do not view aircraft carriers as the credible combat force they were in 1996, diminishing their "presence value."[12] Even if the Chinese are wrong, it does not matter; Chinese perceptions—not reality—determine the carrier's effectiveness as a deterrent.[13] The Flex Fleet's superior warfighting abilities give it more credibility than the carrier-centric model, making it a structure better suited to execute peacetime presence missions.

The Flex Fleet is built to maximize fleet presence, not individual ship presence. Individually, each new Flex Fleet platform is less impressive than *Ford*-class aircraft carriers and *Arleigh Burke*–class destroyers. However, when these smaller platforms are alloyed with the larger destroyers and carriers, they can provide a much stronger message than that of today's Navy. President Theodore Roosevelt said that a strong Navy is "the surest guaranty of peace."[14] The Flex Fleet, with more ships, improved ability to tailor ships to missions, and more warfighting credibility, is a better "guaranty of peace" than the U.S. Navy as it is structured today.

Reconnaissance

The fleet's ability to find and track the enemy above, on, and below the ocean's surface is pivotal. It directly affects all other mission areas, from anti-surface warfare to maritime interdiction operations. Today, the U.S. Navy has an unmatched ability to find and track adversaries. Using tools such as the E-2 Hawkeye, SPY-1 radar, satellite networks, and cyber capabilities, it has achieved unprecedented battlespace awareness. However, the U.S. Navy's reconnaissance abilities are far from complete and can be improved.

The Flex Fleet would be even better at reconnaissance than today's force structure. With a 42 percent increase in battle force ships over the Navy's planned inventory, the Flex Fleet would have more ships searching for the enemy. No longer required to primarily concentrate in carrier strike groups, that increased number of ships could be more effectively distributed to achieve improved geographic coverage. Furthermore, the Flex Fleet has 56 percent more ships able to operate aircraft than the U.S. Navy plans on having in 2049, meaning that numerically superior and dispersed force can further expand its search from the air using helicopters and UAVs. The Flex Fleet will also be better able to operate in the littorals, since ships like Steel-class corvettes have smaller signatures and can be more easily risked than large, expensive ships like *Arleigh Burke*–class destroyers. That presence in the littorals will better enable the Flex Fleet to overcome the clutter of civilian shipping and the nearby coastline to detect and identify the enemy. The Flex Fleet improves on the U.S. Navy's reconnaissance foundation by leveraging more ships better spread out carrying more aircraft to achieve improved geographic and spectrum coverage. The Flex Fleet strives to achieve the scouting gauge, giving commanders early knowledge of the enemy and thus the option to attack effectively first or disengage.

The Flex Fleet does suffer from a reduced ability to use the airborne early warning E-2 Hawkeye, as it cannot operate from Bronze-class light aircraft carriers. However, their loss is more than mitigated by increased numbers of platforms launching aircraft and UAVs in the Flex Fleet. Additionally, E-2s can still be operated from the remaining *Nimitz*-class and *Ford*-class carriers, as well as from land, if operating bases are available. Under the Flex Fleet plan, those CVNs will remain in service into the 2070s, giving the Navy decades to develop alternative airborne early warning options, such as Boeing's proposed airborne early warning variant of the V-22 Osprey tilt-rotor aircraft.[15] In the meantime, Bronze-class carriers can use helicopters or UAVs in an airborne early warning role, as other nations do.[16]

The decision to reduce the fleet's reliance on the E-2 is similar to the decision to reduce the U.S. Navy's reliance on the aircraft carrier. The E2 is, individually, better at reconnaissance than any other aircraft type. However, a reconnaissance network primarily reliant on a small number of E-2s is not as effective as one like the Flex Fleet. The E-2 is a highly capable platform, but no aircraft is so important that it should dictate the entire structure of the U.S. Navy.

Anti-Surface Warfare

The Flex Fleet would also outperform the U.S. Navy in anti-surface warfare because it is built to capitalize on the opportunities identified in Chapter 1.

First, the Flex Fleet embraces the Age of the Missile by shifting many of the carrier's offensive operations to missiles. The Flex Fleet has approximately 2,900 more missile cells than the Navy's planned fleet, a 16 percent increase.

Furthermore, the Flex Fleet can more rapidly reload those cells because each individual ship tends to be less important to the overall fleet than the top-heavy structure of the U.S. Navy today. For example, when a Brass-class missile ship launches its salvo of missiles and withdraws to port to reload, all the fleet loses is the ship's already expended arsenal. In contrast, when an *Arleigh Burke* destroyer withdraws to reload, the Navy also loses its many capabilities in mission areas such as anti-air defense, antisubmarine warfare, and command and control. As a result, a Brass-class missile ship could immediately depart to reload its missiles whereas an *Arleigh Burke* destroyer may be delayed to perform other operations. Thus, the Flex Fleet not only has more missiles, but can reload them faster as well.

Second, the Flex Fleet better utilizes the Navy's networks by having more sensors feeding it and more shooters benefiting from it. With more platforms operating more aircraft, the network will be able to investigate more detections and track targets for longer periods of time. Without having to always link back to a central node—the aircraft carrier—those platforms can generate their own "local combat-information networks," which are "essential to achieving localized battlespace awareness," according to the three surface warfare admirals who introduced the concept of Distributed Lethality.[17] That enables the fleet to share data and fight even when long-range communications are lost, reducing the burden on the aircraft carrier to be the nexus of all fleet communications. In modern warfare, the network matters most, and the Flex Fleet would have a stronger network than that which is possible with a carrier-centric fleet.[18]

Third, the Flex Fleet diversifies the kill chain by enabling numerous platforms to destroy the enemy. Compared to the fleet the U.S. Navy intends on having in 2049, the Flex Fleet has 86 percent more missile-shooting surface ships, with many of those ships able to launch aircraft or UAVs that can also attack enemy ships. Instead of having to primarily rely on the carrier air wing and submarines, ships like corvettes, frigates, and missile arsenal ships can launch their own credible attacks. They can conduct those attacks independently, provide targeting data to missile arsenal ships, or launch aircraft to carry out strikes. Destroyers, with a reduced need to serve in a high-value unit protection posture, can shift to the offensive and bring their impressive capabilities to bear. Supporting all this the carrier air wing can conduct its own anti-shipping attacks or focus on providing air cover, similar to how the British used their carriers in the Falklands Conflict.[19] The Flex Fleet benefits from missiles and strong networks to provide commanders the flexibility and kill chain diversification they do not have in today's carrier-centric structure.

In the Flex Fleet, those commanders have a variety of platforms enabling them to tailor their force to the situation and the degree of sea control attained. For example, early in a conflict, submarines and Steel-class corvettes, with their small signatures, maneuverability, shallow draft, and heavy armament, could probe the littorals. Acting as a first wave for the fleet, the corvettes can search for the enemy using their radars, electronic support measures, and UAVs, following any detection with a missile attack. They would be powerful enough to hold their own in combat but not so valuable that their destruction would be crippling.

As the submarines and corvettes weakened the enemy and provided the tactical picture to the rest of the fleet, ships like *Constellation*-class frigates and Brass-class missile arsenal ships could join the fight. Even with the Brass-class ships hundreds of miles away, they can launch anti-ship missiles using targeting data from the corvettes, frigates, submarines, or aircraft. The light carriers and accompanying destroyers could provide air cover and could conduct ASUW missions as well. By focusing on fleet air defense instead of land-attack missions, the light carriers can operate farther out to sea to maximize their protection.

Distributing the force to more ships also makes its total destruction less likely, mitigates the repercussions of a single enemy missile hitting a friendly target, and strains the adversary's resources. To illustrate this point, compare the number of enemy ASCMs hypothetically required to sink an *Arleigh Burke*-class destroyer or four Steel-class corvettes.[20] Assume the probability of detecting and tracking the destroyer or a single corvette is 50 percent, although the corvette has a much smaller signature. Thus, the probability of finding the destroyer is 50 percent, whereas the probability of locating all four of the corvettes is 6.25 percent. Assume that three ASCM hits are required to disable the destroyer—likely an overly hopeful estimate—and only one to destroy the corvette. Further, assume that each ASCM has a 50 percent probability of tracking and hitting either target; missiles are more likely to detect and hit the destroyer due to its significantly larger radar cross section and infrared signature, but its hard-kill systems are more capable than those of the corvette, possibly making up the difference.[21]

In this scenario, if the enemy wants an 80 percent confidence of destroying the opposing force, it would have to launch eight missiles against the destroyer. However, to have an 80 percent probability of destroying the corvette flotilla, the enemy would require a 94.5 percent likelihood of destroying each individual corvette. Therefore the enemy would have to fire five missiles per corvette, or twenty missiles total to achieve the same 80 percent confidence of destruction. The enemy would have to fire 50 percent more missiles than the number required to neutralize or sink the destroyer to defeat the flotilla of just four corvettes. The corvette force is harder to find and requires more missiles to be destroyed.

The Flex Fleet's improved ability to attack effectively first, coupled with its improved survivability against enemy surface forces, enable it to outperform the carrier-centric Navy in the anti-surface warfare mission area.

Antisubmarine Warfare

Antisubmarine warfare is a "team sport" requiring numerous highly trained, coordinated platforms piecing together infrequent clues to eventually find and destroy or deter the enemy submarine.[22] In this sport the quality of the searcher matters, but the quantity of searchers is an important factor as well. Lots of air, surface, and subsurface platforms are needed to cover vast ocean expanses and conduct a wide-area search for submarines, then track and destroy them. A surface fleet with *Arleigh Burke*–class destroyers, *Constellation*-class frigates, and Steel-class corvettes, as well as their embarked aircraft, will have both the quality and quantity of searchers needed for this mission and be more effective than a surface fleet primarily reliant on just destroyers. The Flex Fleet, with a 58 percent increase in antisubmarine warfare-capable surface ships, will have better odds of finding and destroying submarines.

In today's fleet structure, antisubmarine aircraft embarked on the carrier are a defensive system in case of nearby threats. However, if submarines are within the embarked helicopters' range, then those submarines' anti-ship missiles are also likely within range of the carrier. The Flex Fleet is designed to go on the offensive in the undersea realm, spreading its numerically superior fleet to find and destroy those submarines before they can close to attack high-value units. If submarines do close within missile range, it is likely too late for those helicopters to have a major impact.

By acknowledging this reality, the Flex Fleet can also strengthen carriers' air wings. Today, approximately 8 percent of air wings are antisubmarine helicopters, a large investment and opportunity cost.[23] In contrast, the Flex Fleet does not attempt to force the carrier to accomplish a mission it is ill-suited for. Instead, the Bronze-class light carriers devote almost all their air wing space to tactical fighters to perform missions the carrier excels at, such as anti-surface warfare and fleet air defense, and not attempt missions for which it is poorly suited. By having more antisubmarine warfare-capable platforms and shifting more of them to the offensive, the Flex Fleet is better poised to control the undersea area of operations than the carrier-centric Navy.

Integrated Air and Missile Defense

The Flex Fleet is a more effective option than today's carrier-centric fleet for the integrated air and missile defense (IAMD) mission because it focuses on preventing those attacks, is better suited to defeat them when they occur, and can continue to fight even if some enemy attacks are successful. Most importantly the Flex Fleet is built to acknowledge the reality that missile defenses are generally ineffective. Instead of pouring billions of dollars into futile efforts to overcome that truth, the Flex Fleet shifts its resources and efforts to areas it can realistically affect, yielding a better overall result.

Missile defenses are generally ineffective because they always start at a disadvantage. The attacker has time, physics, and numbers on his side. As Israeli admiral Yedidia Ya'ari summarized in a famous 1995 *Naval War College Review* article,

> Where missiles are concerned, the contest between the offense and defense is marked by a serious differential in starting points. In practical terms, the offense has a huge and nearly motionless target to hit and needs to hit it only once. One large missile warhead is equivalent to something like five or ten direct hits by a sixteen-inch gun. The defense,

on the other hand, is required to intercept an extremely fast and quite agile flying object, sometimes hardly detectable in the various phases of its trajectory, which can be launched from any operational dimension and often—for design purposes, *every time*—completely by surprise. The defense must deal with a weapon that can perform deceptive terminal maneuvers intended to outmaneuver hard-kill means (those attempting to actually destroy the missile); with a weapon equipped with any, or a combination, of a variety of guidance systems and homing devices designed to outperform a ship's "soft-kill" protective measures (which attempt, actively or passively, to cause the missile to miss); with a weapon that can be launched in salvos on multiple approach paths to saturate countermeasures of whatever kind.

Above all, the defense must constantly perform without error and without defect in an electronic environment so densely charged and a tactical situation so cluttered that they cannot be fully simulated. Uncertainties regarding the actual performance of defensive suites in a full-blown modern engagement are a cause for concern. Even limited experience has established, however, that whereas for the offense a mistake or malfunction means the loss of a missile, for the defense it means at least the disablement, and probably the loss, of a ship.[24]

Missile defense systems' historical performance has validated those inherent disadvantages. Interceptors have shot down only a single anti-ship missile in the more than fifty years of their existence, when HMS *Gloucester* destroyed an old, slow Silkworm missile during Operation Desert Shield. It is possible that USS *Mason* destroyed Houthi missiles near Yemen in 2016, but it was never confirmed and appears unlikely.[25] Of all anti-ship missile attacks between 1967 and 1994, attacks on warships that could have defended themselves but did not were successful 68 percent of the time, and attacks on warships that executed defensive measures were successful 26 percent of the time.[26] Many of these attacks were simple single missile salvos, a best-case scenario for defenders. This performance led the "New Navy Fighting Machine" authors to conclude in their report for the secretary of defense's Office of Net Assessment, "It is not certain that defense with surface-to-air missiles was ever as good as is commonly believed."[27]

Acknowledging this reality, the Flex Fleet shifts its resources and attention to what it can better affect: the early destruction of the enemy, thus negating the need to interdict incoming air and missile attacks. The Flex Fleet does not abandon missile defenses, as they will remain key to the fleet's operations. Instead, the Flex Fleet strives to "attack effectively first" even more than today's fleet does, knowing that it is the best defense.

The U.S. Navy today obviously also strives to prevent the enemy from launching attacks. However, its structure prevents it from accomplishing this task as well as the Flex Fleet. As shown above, the Flex Fleet is more effective on the offense because of its improved reconnaissance, anti-surface warfare, and antisubmarine warfare capabilities. Furthermore, the U.S. Navy has poured so much into its aircraft carriers—$12.8 billion, an enormous crew of five thousand, and the symbol of American might—that it has no choice but to pour billions into defensive systems for such a vital asset. Yet with a fixed budget, every dollar devoted to protecting the carrier is a dollar not spent on attacking effective first. The Flex Fleet outperforms the carrier-centric fleet in integrated air and missile defense because it is better at preventing the need for integrated air and missile defense.

When the enemy is able to launch an air or missile attack, the Flex Fleet is again superior to the carrier-centric Navy because of its superior performance at each step in the kill chain. The smaller platforms of the Flex Fleet would have reduced signatures, and with their increased numbers and geographic spread they would be harder to track than a small number of large carrier strike groups. With better reconnaissance, the Flex Fleet can identify an incoming attack earlier, and with tactical fighters like the F-35 spread out onto more light carriers there will be more nearby options to intercept that attack. Those F-35s and light carriers, with a smaller mission set than today's large carriers, will be better focused on the air defense mission. The Flex Fleet would also have a better spread of soft-kill and hard-kill defensive systems installed on smaller, more maneuverable targets that are harder to hit. As Adm. Arthur Cebrowski and Stuart Johnson wrote in their "Alternative Fleet Architecture Design" report, "Survivability can increase as size decreases. Small platforms can be designed with greater speed, maneuverability, and a smaller signature than large ships. This allows them to elude detection, tracking, and strike by an enemy rather than relying on thickness of armor for survivability. Moreover, their greater numbers makes [sic] it harder for an enemy to establish and maintain track on the total force and to determine what to attack to get the maximum payoff."[28]

Finally, the Flex Fleet is also superior because it can sustain the ship losses that will occur when those air and missile defenses inevitably fail. If the enemy can sink or damage carriers of the *Nimitz* or *Ford* class today, the results would likely be catastrophic. Roughly 10 percent of the class would be gone or incapacitated in a single action, the strike group would lose its primary method of delivering firepower, and the image of a burning aircraft carrier would have severe political and diplomatic consequences. Centralizing so much power on a single ship is economically feasible and tactically effective only if it can operate. When it is lost, all those capabilities are gone as well, weakening the entire fleet. In contrast, no ship is so vital to the Flex Fleet's success that its loss would be disastrous. The Flex Fleet is built with the knowledge that it will not always be able to attack first, that its defenses will periodically fail, and that some ships will sink. It prepares for that future by spreading out its capabilities to as many platforms as possible so that the loss of a single ship does not handicap the entire network.

The Flex Fleet can better prevent attacks, has a better defense when attacked, and can better sustain losses if attacks succeed. It is grounded in reality and respects the lessons of naval history and modern technology. It does not force huge portions of the fleet to futilely attempt the impossible: a perfect defense for a priceless ship.

Strike Warfare

If the fleet can prevail in mission areas such as anti-surface warfare and antisubmarine warfare, it can use that sea control to project power ashore, primarily using aircraft-delivered weapons and ship- and submarine-launched missiles. Strike warfare is an exceptionally important mission area and one the aircraft carrier has historically excelled at.

The *Ford* class of carriers was designed to be the world's premier strike platform by having an unmatched sortie generation rate (SGR), which the Navy's operational requirements for the class defined as "the defining measure of the supported combat power of an aircraft carrier. It is the measure that includes the ability to launch, recover, service, load and prepare the aircraft in all ways for the succeeding mission. This includes requirements for both the flight deck systems attendant in aircraft operations and the sustainability requirements for magazine and fuel stowage."[29]

In benign environments, where the enemy has little ability to contest the seas or air, the aircraft carrier is undoubtedly the best platform for power projection because of its high SGR. Its powerful air wing, large weapons inventory, ability to resupply at sea, and capable maintenance infrastructure enable it to sustain large numbers of lethal attacks over long periods of time. No other platform or fleet structure, including the Flex Fleet, can match its SGR and thus volume of attacks on targets ashore. The Flex Fleet's smaller Bronze-class light carriers would also be able to generate a large number of air strikes, but at reduced rates and over shorter periods of times compared to *Nimitz*- and *Ford*-class carriers. The Flex Fleet's primary strike platforms, the Brass-class missile arsenal ships, can generate large missile attacks but then must withdraw to port and reload before launching additional attacks.

The aircraft carrier has repeatedly demonstrated this impressive performance throughout its history. During the Vietnam War, aircraft flying from carriers flew more than half the strike missions against targets in North Vietnam.[30] In the initial phase of Operation Enduring Freedom, although U.S. Air Force bombers featured the preponderance of munitions, U.S. Navy aircraft flying from carriers executed 75 percent of all strike missions.[31] In a benign environment, where the enemy cannot effectively contest the U.S. Navy's sea control or air superiority, the carrier-centric model is the best option for power projection.

However, in a contested environment, a credible enemy would challenge either fleet structure's ability to attack targets ashore. That enemy, likely with areal denial weapons and integrated air and missile defenses, may push the U.S. Navy outside of the range of its aircraft and missiles. This would force the fleet to a raiding posture, in which it would sail within range of enemy weapons, launch an attack, and withdraw.

In that raiding posture, both fleet structures would be able to conduct fewer attacks than in a benign environment. However, several factors would negatively affect the carrier Navy more than the Flex Fleet, reducing both the frequency and intensity of its attacks. The carrier would likely need to sail closer to its target than the Flex Fleet because manned aircraft, with limited ability to conduct in-flight refueling near credible enemies, generally have shorter ranges than missiles which only need to fly one way. For example, the F35 has a combat radius of approximately 750 nm, compared to the Tomahawk missile's range of roughly 900 nm, depending on each variant.[32] Once within range of the enemy's defenses, the carrier would need to stay there longer to recover its air wing after the attack, whereas Flex Fleet platforms like *Brass* class missile arsenal ships could withdraw as soon as they launch their missiles. Most importantly, enemy air and sea defenses negates the carrier's greatest strength, its ability to sit off a coast and use its high SGR to create large volumes of continuous air strikes. Those defenses would force the carrier into a posture that the Flex Fleet is already designed to employ: attack, withdraw, reposition, and reattack. Overall, both the carrier-centric fleet and the Flex Fleet would be able to conduct less frequent attacks against a peer adversary than against a weak opponent. However, a navy relying on the aircraft carrier would suffer a much larger reduction in frequency of attacks than the Flex Fleet.

A competent enemy with integrated air and missile defenses would also affect the strength of those attacks. The carrier air wing would need to devote part of its air wing to serve as fighter escorts, conduct electronic warfare, and provide tanking services, reducing the number of aircraft available to attack enemy ground targets. The Flex Fleet's missile shooters would not need to conduct those support functions, meaning their missile arsenals could continue to be primarily devoted to offensive strikes. To maintain the F-35's stealth features when attempting to penetrate enemy air defenses, it could only carry weapons in its internal bays, thus reducing its payload by 74 percent.[33] The Flex Fleet's missile payloads, on the other hand, would remain unchanged. Finally, attacks coming from aircraft carriers will likely have a degree of planned conservatism to reduce the likelihood of losing manned aircraft, which would result in important tactical, financial, and political repercussions, another problem the Flex Fleet mitigates by relying on missiles.

The RAND Corporation studied many of these problems in its 2017 congressionally-directed study on aircraft carriers. RAND's report stated,

"Although there are operational circumstances in which the ability to generate large numbers of sorties rapidly would be important, for the most-stressing scenarios, this will likely not be the case. The most-stressing scenarios—those involving a near peer with significant defensive capabilities—will likely not allow the [carrier strike group] to close the target area until after significant suppression of the enemy's air defense and counter-maritime capabilities have been diminished."

The report concluded that designing a carrier with a high SGR "as a principal and overarching characteristic is seeking a capability that is highly relevant in only a very narrow set of circumstances."[34] Captain Hendrix stated frankly, "Those who have argued or will argue for sortie generation proceed from a false premise that both the oceans and the air above them will remain accessible and permissible to U.S. naval assets. This is an assumption that is neither borne out by history nor supported by strategic logic."[35]

Overall, the enemy's warfighting skill determines whether the carrier-centric fleet or Flex Fleet is the best option for strike warfare. Against foes like Iraq, Serbia, or Afghanistan, with little ability to defend against a modern military, the carrier's high SGR makes it the best choice. However, against peer adversaries such as China and Russia, the carrier's growing susceptibility to area denial weapons and its air wing's short range and limited ability to penetrate robust air defenses negates the benefits of a high SGR and seriously challenges its ability to conduct strike warfare. As defense analyst Robert Haddick wrote in *Fire on the Water*, "In a conflict with China, the U.S. Navy as it currently exists is largely out of the power projection business."[36] In comparison, the Flex Fleet can attack at greater ranges with more platforms launching more missiles better able to penetrate defenses. When the threat is contained, the Flex Fleet can then augment that force with Bronze-class light carriers to increase its strike volume. In contested fights, the Flex Fleet is the better option for strike warfare.

AN AIR WING-TO-AIR WING COMPARISON

In 2015 Captain Hendrix published "Retreat from Range: The Rise and Fall of Carrier Aviation."[37] He identified five characteristics key to evaluating carrier air wings throughout history: mass of platforms, range, payload capacity, low observability, and persistence. That framework enables a comparison of the air wings of the carrier-centric fleet and the Flex Fleet, providing more insight into which is the best option for the U.S. Navy now and in the future.

Captain Hendrix defined mass as the size of the air wing and how many aircraft could engage a target at any time, while range and payload capacity defined how far from the carrier they could operate and what munitions they could deliver. Later in the development of carrier aviation, low observability defined the air wing's stealth features, and persistence characterized the air wing's ability to stay on station for extended periods of time.

Comparing the air wings of a single *Nimitz*-class or *Ford*-class CVN or a single Bronze-class CVL, there is no contest. The CVN air wing has far superior mass, with approximately forty-four tactical aircraft and twenty-five other aircraft for support missions. Its jet aircraft are conventionally launched, meaning they can takeoff with more fuel and more munitions than STOVL aircraft, enabling the CVN air wing to outperform that of the CVL in both range and payload capacity. While the air wings' stealth features are likely similar, the CVN air wing has more aircraft to support limited tanking services, meaning it is also the more persistent option. Most importantly, the CVL air wing does not include the E-2 Hawkeye, limiting its airborne early warning capabilities. In a one-to-one comparison it is not surprising that the CVN air wing outperforms that of a CVL. If the Navy replaces its eleven CVNs with eleven CVLs, it will result in a markedly weaker fleet.

When comparing the combined airpower of all CVN air wings with that of the sixteen CVL air wings of the proposed Flex Fleet, the comparison becomes closer. While the characteristics of range, low observability, and persistence are for the most part unchanged, the total number of aircraft employed by all CVNs versus all CVLs—defining the mass and payload capacity—is much closer, with the Flex Fleet operating approximately 9 percent fewer F-35s overall. Additionally, the loss or unavailability of one CVN has a much larger effect on today's Navy than a CVL would have on the Flex Fleet. The Flex Fleet air wings can attack from more carriers, giving them the ability to fly from more vectors and cover more area, mitigating the range shortfalls. However, using Captain Hendrix' framework, the air wings of the Navy's current structure still outperform those of the Flex Fleet. If all the Navy does is stop building CVNs and use that money to purchase sixteen CVLs, it will again result in a somewhat weaker fleet.

The correct comparison is not between individual ships or two classes of ships, but instead between entire fleet structures. In a 2020 discussion of previous Pentagon studies that concluded that larger nuclear-powered carriers were the better option compared to light carriers, Chief of Naval Operations Adm. Michael Gilday said those reports tried to make "an apples-to-apples comparison" that "lead to a fait accompli that a smaller carrier just does not compete with a super carrier. I think that's a false choice."[38] The Flex Fleet is the solution to that analysis problem; the comparison is not CVN versus CVL, but the airpower of the entire carrier-centric fleet versus that of the entire Flex Fleet, beyond what the carriers offer. In that context the Flex Fleet has more mass in the air, with its small reduction in F-35s offset by increased numbers of UAVs and missiles launched from more platforms. The Flex Fleet's 16 percent increase in missile cells and 86 percent increase in missile shooters enables it to outrange today's fleet structure, heavily reliant on manned aircraft with ranges typically less than those of even older missiles like the Tomahawk. Those manned aircraft, even with stealth features, likely do not have a radar or visual cross section as small as that of a missile, again swinging the balance in favor of the Flex Fleet. Finally, with a 56 percent increase in ships that can launch aircraft, including long-range UAVs, the Flex Fleet will have more persistent airpower. The Flex Fleet still makes heavy use of manned aircraft, but its increased reliance on missiles and unmanned platforms operating from a distributed fleet means its airpower has more mass, range, payload capacity, stealth, and persistence than that of the planned U.S. fleet. Figure 3.2 summarizes these air wing comparisons.

In an article advocating for distributing naval airpower, Capt. Wayne Hughes wrote, "In recent years, awareness of the need for a bigger navy has grown, but most solutions have included buying more of the same kind of warships or changing the mix of existing designs. This is like shuffling deck chairs on the *Titanic*—with the CVN as the only deck chair design for sea-based air operations."[39] The Flex Fleet is one alternative to merely "shuffling deck

FIGURE 3.2. U.S. Navy and Flex Fleet Airpower Comparisons

Air Wing Characteristic	1 CVN Air Wing vs. 1 CVL Air Wing	F10 CVN Air Wings vs. 16 CVL Air Wings	USN Airpower vs. Flex Fleet Airpower
Mass	Winner: CVN CVN operates approximately 44 F 35C, plus support aircraft, vs. just 30 F 35B on a CVL	Winner: CVN 10 CVNs operate approximately 10% more F-35 than 16 CVLs	Winner: Flex Fleet Small reduction in F-35 offset by 16% increase in missile cells, 86% increase in surface warship missile shooters, increase in UAVs
Range	Winner: CVN Conventionally launched F-35C has greater range than STOVL F-35B	Winner: CVN Conventionally launched F-35C has greater range than STOVL F-35B	Winner: Flex Fleet F-35B range reduction offset by significantly increased use of missiles with ranges greater than F-35C
Capacity	Winner: CVN Conventionally launched F-35C has greater payload capacity than STOVL F-35B	Winner: CVN Conventionally launched F-35C has greater payload capacity than STOVL F-35B	Winner: Flex Fleet F-35B payload capacity reduction offset by increased use of missiles
Observability	Tie	Tie	Winner: Flex Fleet Flex Fleet increases its reliance on missiles, with a smaller radar cross section
Persistence	Winner: CVN More aircraft provide more tanking options	Winner: CVN More aircraft provide more tanking options	Winner: Flex Fleet UAVs from corvettes/ frigates and increased number of missiles provide more persistent options
Overall	Winner: CVN 1 CVN air wing outperforms 1 CVL air wing	Winner: CVN 10 CVN air wings outperform 16 CVL air wings	Winner: Flex Fleet Through combination of conventional airpower (CVLs), increased use of UAVs (corvettes, frigates), and a large increase in the use of missiles, the Flex Fleet outperforms the carrier-centric fleet

chairs," and through the use of conventional aircraft, STOVL aircraft, more unmanned platforms operating from more warships, and increased numbers of missiles and missile platforms, it strives to achieve a "more distributable air force instead of concentrating so much value in 10 or 11 nuclear-powered aircraft carriers."[40]

UNINTENDED BENEFITS AND CONSEQUENCES

The Flex Fleet's primary goal is to outperform today's U.S. Navy in its principal mission areas. It strives to show that it is possible to design a technically and financially realistic fleet that benefits from aircraft carriers without being wholly dependent on them. While the peacetime and warfare mission areas are the focus, there are secondary consequences that affect whether the Flex Fleet or the carrier-centric Navy is the best structure going forward. Those consequences can be binned into the categories of construction and maintenance of ships, their crews and training, and their operational costs.

The Flex Fleet will have important effects on the nation's shipyards. Today, the U.S. Navy's four public shipyards, augmented by private companies, are strapped to complete all required maintenance periods on time and within a reasonable budget. The small number of large private shipyards limits overall capacity and reduces competition, thus raising costs. For example, between 2015 and 2019, just 25 percent of nuclear-powered aircraft carriers and submarines completed their maintenance periods on schedule, resulting in a combined 7,424 days of delay.[41] These delays lead to massive cost overruns and prevent the start of work on other ships. Submarine idle time is the time between when they can no longer operate because of lapsed safety certifications and when a shipyard spot becomes available and they can start their maintenance periods; between 2015 and 2019, submarines sat idle for 2,796 days, meaning the U.S. Navy lost the equivalent of almost eight years of operational submarine time. It is a problem that is only getting worse, as the annual count of idle time increased by 919 percent over those five years.[42]

The Flex Fleet, reliant on ships that tend to be smaller and less complex, mitigates many of those problems. Ships like Steel-class corvettes and Brass-class arsenal ships, with smaller hulls and simpler designs than large capital ships, are easier to build and maintain. As a result, more shipyards across the country can work on them, alleviating the burden on the few shipyards able to work on the largest platforms. If the Navy has more shipyards to choose from, that increases overall capacity and injects more competition into the industry, helping to reduce costs, delivery schedules, and idle time. Additionally, more ships of the Flex Fleet are conventionally powered, meaning they do not incur the high maintenance costs and lengthy overhaul periods associated with nuclear propulsion plants. The Flex Fleet will not solve the Navy's construction and maintenance woes, but through design simplification and shipyard diversification, it can help reduce them.

The Flex Fleet's effects on shipyards are overall positive, but there is one primary drawback. At present only one U.S. shipyard—in Newport News, Virginia—can build nuclear-powered aircraft carriers.[43] If that shipyard stops building Ford-class carriers, it would be prohibitively expensive for it to keep employing those thousands of skilled workers without work for them. As a result, the company would likely lay off large portions of the only workforce in the United States that can perform nuclear work on that scale. Under the Flex Fleet plan, the U.S. Navy will stop ordering Ford-class carriers after USS Enterprise is delivered in 2028, meaning that the United States would no longer be able to build nuclear-powered aircraft carriers without a massive infusion of money and time to regain the necessary people, expertise, and equipment.

This consequence of the Flex Fleet's shipbuilding plan is unintended but acceptable. The development of ironclads in the nineteenth century meant that the U.S. Navy eventually lost proficiency building wooden sailing ships. The development of aircraft carriers in the twentieth century meant the U.S. Navy eventually lost proficiency building battleships. The introduction of alternative fleet structures in the twenty-first century means the United States will eventually lose proficiency building nuclear-powered carriers. In each case, it may have seemed radical and alarming to depart from reliable, well-known ship designs, but part of boldly seizing future opportunities is letting go of the past. Even if initial Flex Fleet designs are flawed, the solution will be to improve those designs, not revert to old ones by restarting CVN construction. As Capt. Robert Rubel wrote, "It does not seem reasonable to presume that the strategic future of the United States hinges on a few thousand shipyard workers in Virginia."[44]

Once the ships are built, the Flex Fleet also has advantages over today's fleet structure in manning those ships. Recruiting and retaining sailors and officers to operate the Navy's ships is always a challenge, due in part to the service's high standards and challenging lifestyle. For example in February 2020 the Navy had approximately 9,000 unfilled billets on ships, falling 6 percent short of its manning target.[45] Shifting to a fleet with fewer aircraft carriers could alleviate this problem. In a Naval Postgraduate School thesis for Captain Hughes, then-lieutenant Juan Carrasco compared the manning requirements for the Navy's 280 ships in 2009 and the "New Navy Fighting Machine," a hypothetical 677-ship fleet that incorporated smaller platforms like those used in the Flex Fleet. Carrasco determined that the "New Navy Fighting Machine," despite having more than double the number of ships of the U.S. Navy in 2009, required 10 percent fewer people to man them. He also determined that the Navy's aircraft carriers, composing just 4 percent of the fleet's warships at the time, required 47 percent of personnel afloat. In a similar comparison shown in Figure 3.3, the Flex Fleet would also require 10 percent fewer people at sea than the fleet the Navy is planning to have in 2049. Not only would the Flex Fleet have fewer billets to fill, but it may have an improved ability to retain personnel to fill them. The Flex Fleet's increased number of small ships means that more officers and enlisted sailors will rise to command and leadership roles earlier in their careers, likely improving recruiting and retention of the professionals the Navy needs.[46]

Just as the Flex Fleet's design provides advantages in ship maintenance and personnel, it also enables better warfare training on those ships. Today, platforms like *Arleigh Burke*–class destroyers must execute a wide array of missions because there are limited smaller platforms able to share the load. However, being forced to execute numerous missions reduces the crew's proficiency and expertise in any one area. As Adm. Jonathan Greenert wrote when he was Chief of Naval Operations, *Arleigh Burke* destroyers' large set of required operations "show the limitation of a highly integrated luxury car platform. While the ship, aircraft, and crew might flex to new or different missions, it does so at a cost. Destroyer crews are challenged to maintain proficiency in core missions such as [antisubmarine warfare], [anti-surface warfare], and [integrated air and missile defense] when engaged in months-long counterpiracy operations."[47]

Cdr. Bryan McGrath also discussed the challenge of maintaining crew proficiency and expertise in key mission areas such as antisubmarine warfare with competing priorities. In congressional testimony he said, "And while my ship the USS *Buckley* [DDG 84] had an outstanding sonar suite—far better

FIGURE 3.3. U.S. Navy and Flex Fleet Manning Comparison

	Ship Type	USN Target Inventory in 2049 (Ships)	Flex Fleet Target Inventory in 2049 (Ships)	Crew Size per Ship
Modified in the Flex Fleet	Corvettes Steel Class (DDC)	0	110	45
	Small Surface Combatants *Freedom* Class (LCS) *Independence* Class (LCS) *Constellation* Class (FFG)	50	81	98
	Missile Arsenal Ships Brass Class (MAS)	0	28	98
	Large Surface Combatants *Arleigh Burke* Class (DDG) and Replacement	108	75	329
	Aircraft Carriers *Ford* Class (CVN)	10	0	3,100
	Light Aircraft Carriers Bronze Class (CVL)	0	16	1,204
	Combat Logistics Ships (CLF)	31	40	109
	Large Payload Submarines *Columbia* Class (SSGN)	3	0	159
Unchanged	Amphibious Warfare Ships	35	35	700
	Ballistic Missile Submarines *Columbia* Class (SSBN)	12	12	159
	Attack Submarines *Virginia* Class (SSN) and Replacement	67	67	132
	Support Ships	39	39	50
Flex Fleet Net Decrease in At-Sea Manning Requirements				-11,357 Personnel
Flex Fleet Net Decrease in At-Sea Manning Requirements				-10%

Table Notes

1. Crew sizes for existing platforms from the U.S. Navy Fact Files. Does not include carrier air wings. Source: "U.S. Navy Fact File," U.S. Navy, accessed August 29, 2020, https://www.navy.mil/Resources/Fact-Files/.
2. Crew sizes for multiple classes composing one category of ship (i.e., multiple kinds of amphibious ships) determined by using approximate average of all ship classes' crew size.
3. Crew sizes for new ship classes estimated using comparable ships of similar size and mission. For example, the Visby-class frigate crew size of approximately forty-five people was used for Steel-class corvettes. Source: Chris Summers, "Stealth Ships Steam Ahead," BBC News, June 10, 2004, http://news.bbc.co.uk/2/hi/technology/3724219.stm.

than which I had on my first ship in 1987—years of decline in surface force ASW proficiency due to the mission having been deemphasized resulted in an unshakeable conviction that if I had to face a submarine threat, I would rather have done it on the old, loud frigate with the highly proficient team than on my new destroyer whose complement of sonar technicians had declined along with their proficiency."[48]

The Flex Fleet would minimize these oversubscription problems by focusing its platforms on fewer missions. Steel-class corvettes could achieve high levels of expertise in anti-surface warfare, Brass-class missile arsenal ships could primarily focus on strike warfare, and *Constellation*-class frigates could lead the fleet in antisubmarine warfare. Individual ship classes' mission specialization would alleviate some of the burdens on *Arleigh Burke*–class destroyers and *Nimitz*- and *Ford*-class carriers, allowing them to primarily train for missions only they can execute, such as theater antiair defense. In the Flex Fleet, individual ships' training and mission requirements are less broad, but knowledge of assigned missions is deeper, yielding a more tactically skilled fleet overall.

Finally, in addition to easing the Navy's maintenance, manning, and training burdens, the Flex Fleet has additional operational and financial benefits as well. The shift to a greater reliance on missiles drives most of these benefits. For example, Brass-class arsenal ships have small crews, simplified systems, and VLS cells instead of expensive air wings, nuclear reactors, and complex electromagnetic elevator and catapult systems. As a result, they will have significantly reduced operating and maintenance costs over their service lives than large capital ships, and they will be easier to upgrade when new missiles are developed.[49] When those ships enter battle, they will often prove to be more cost effective than aircraft carriers employing large expensive manned aircraft. Cdr. Phillip Pournelle, formerly of the Department of Defense's Office of Net Assessment, studied the bomb versus missile choice and determined that while aircraft carrying large numbers of bombs are cost-effective in benign environments, standoff weapons such as Tomahawks are a much better choice in even "slightly threatening environments." He discovered that if the risk of losing an F-35, and thus having to pay for a replacement, rises above 2 percent on any sortie, then a missile is the much more cost-effective option.[50] Admiral Greenert also highlighted the financial and operational benefits of shifting missions to missiles:

The ability of a few very-precise standoff weapons to be more efficient and effective than a larger number of less-precise weapons leads to a surprising result. In modern warfare, precision standoff weapons such as Tomahawk or the joint standoff weapon are now more cost-effective in many situations than short-range gravity bombs such as the joint direct attack munition (JDAM). A Tomahawk missile, for example, costs about $1.2 million, while a JDAM is about $30,000. To strike a single target, however, the total training, maintenance, and operations cost to get a manned aircraft close enough to deliver the JDAM is several times higher than the cost of launching a Tomahawk at the same target from a destroyer, submarine or aircraft operating several hundred miles away. That is one of the trends leading us to focus more effort on improving and evolving our standoff sensor and munition payloads.[51]

Whether it is in the construction and maintenance of ships, manning those ships, training the crews, or financing the fleet's operations, the secondary advantages of the Flex Fleet outweigh those of the carrier-centric model. The Flex Fleet would ease problems with overworked shipyards, reduce the number of sailors required to crew the Navy's ships, allow sailors to better focus on training for assigned missions, and cost less money to strike defended targets than today's Navy. The Flex Fleet does introduce new challenges, such as the eventual loss of the ability to build large nuclear-powered aircraft carriers and reduced ability to operate E-2 Hawkeyes. Additionally, relying on a force of more numerous ships will require more supplies, a problem the Flex Fleet addresses by increasing the number of combat logistics force ships by almost 30 percent. Overall, the Flex Fleet's benefits outweigh its unintended consequences, all of which can be overcome without seriously detracting from the fleet's combat abilities.

SAME GAME, NEW PLAYERS

Part I pondered the possibility that the aircraft carrier is growing obsolete. It cited Admiral Nimitz' guidance on the subject: "What determines the obsolescence of a weapon is not the fact that it can be destroyed, but that it can be replaced by another weapon that performs its functions more effectively."[52]

Identifying key opportunities in fleet design and applying them to a hypothetical fleet showed that there are better options than the carrier-centric structure. The Flex Fleet, using the same annual shipbuilding budget and relying on existing technology, outperforms the current U.S. Navy in peacetime presence operations and warfare mission areas and has secondary advantages in ship construction, personnel, training, and operations. Individual ships within the Flex Fleet are all less capable than *Nimitz-* and *Ford-*class carriers, but the Flex Fleet itself is overall more powerful, resilient, and flexible than the planned U.S. Navy force structure. The argument that there is no better alternative to the carrier is no longer valid.

Despite its attractiveness, the Flex Fleet is far from the best replacement for today's Navy. It has many flaws, unanswered questions, and challenges to overcome. It is merely an example of one fleet structure that, by performing the carrier-centric Navy's functions more effectively, fulfills Admiral Nimitz's criteria for proving the carrier's obsolescence. The Flex Fleet concept is not intended to be a perfect blueprint for future Navy force structure. Once the U.S. Navy acknowledges the carrier's waning utility, it can focus its incredible people and awesome resources on designing a new fleet structure that will surely outperform both today's fleet and the Flex Fleet.

When the Navy does realize the carrier's flaws and identifies a replacement fleet, it will likely share some of the Flex Fleet's properties. It will likely be a distributed, flexible fleet able to quickly respond to a variety of peacetime and combat situations. It will likely be built to maximize the fleet's overall capabilities, instead of the capabilities of individual ships. It will seek to capitalize on the Age of the Missile with a networked fleet executing diversified kill chains. That replacement fleet will likely rely on the strengths of numerous ships, each performing roles they can accomplish better than any other platform, rather than relying on extremely expensive multi-mission ships for most operations. That replacement fleet will be built with the knowledge that alloys can have more strength and flexibility than pure metals, and that a fleet of specialized ship classes mixed together is better than a fleet centralized around just eleven increasingly problematic aircraft carriers.

In many respects, today's Navy can be thought of as a baseball team built around a few superstars: *Ford-*class carriers, *Arleigh Burke*–class destroyers, and *Virginia*-class submarines. Building a team around extremely high-caliber players can be a successful strategy, but it has unintended consequences. Investing so much money in just three superstar players means that team may be unable to field the nine total players it needs to be effective, just as today's Navy is unable to build and operate the 355 ships it says it needs to complete its missions. Even if the team can pay for nine players, budgetary constraints will mean those six non-superstars are likely far from effective. Regardless of how good those three superstars are, they only bat once in the lineup, and they can only play one position in the field. Similarly, regardless of how amazing the fleet's carriers, destroyers, and submarines are, they can only do so much and cannot be everywhere at once. Furthermore, that team will quickly fail if just one of its superstars proves ineffective. Whether that superstar's failure is due to injury, a miscalculation of his abilities, or the opposing team's strategy to neutralize him, the result is that the team is now significantly less formidable. Likewise, today's Navy would suffer a large capability reduction if any of its capital ships were unable to perform their functions due to design or maintenance problems, an overestimation of their warfighting abilities, or the enemy's ability to neutralize those ships' power.

Shifting from a roster led by a few superstars to one composed of all capable players will result in an overall stronger team. The opposing team will have to overcome skilled batters throughout the entire lineup, instead of just three superstar hitters. There is no single player that is necessary for victory, and if team management overestimates a player's abilities they can more easily cut him with fewer financial repercussions. The Flex Fleet relies on individually less capable ships to yield a more capable Navy. There is no single ship that, if destroyed, will spell doom for the fleet, and the repercussions of incorrectly predicting a ship class' capabilities will be smaller.

Changing a baseball team—or a fleet structure—can be difficult and temporarily expensive. If today's carrier-centric Navy could continue to effectively perform its missions in the future, then the Navy would be best served by staying with that fleet structure and thus avoid the difficulty associated with change. However, the evidence presented in Part II indicates that the aircraft carrier cannot effectively perform its functions in an increasingly threatening world. Air, surface, and subsurface threats challenge the carrier's ability to survive, never mind its ability to project power and accomplish its mission.

When individuals such as Andrew Marshall and Captain Hughes were originally identifying exciting new opportunities in fleet design several decades ago, the U.S. Navy did not need to adopt them and transform its force structure because there was no serious threat providing the impetus for change. After the Cold War, with the Russian navy essentially bankrupt and the Chinese navy still an obsolete coastal force, the carrier-centric Navy was the best choice for the moment. The U.S. Navy was already clearly superior to any potential adversary, so there seemed little reason to undergo an expensive fleet transformation. The carrier-centric Navy remains a powerful force, but with the rise of the Russian and Chinese militaries, very good is no longer good enough. Thus, while Part I showed *what* the fleet should change, Part II focuses on the carrier's vulnerabilities and opportunity costs to show *why* the fleet should change now.

PART II

RISKS AND OPPORTUNITY COSTS

Battleships dominated naval warfare for about sixty years, and carriers for about the same. Our existing carriers will have plenty to do for the remainder of their operating lives, but a Navy built around these ships will not carry us into the emerging era of warfare any better than did the USS Arizona *into World War II. To procure more large carriers today and expect them to be useful into midcentury is to be blind to reality.*[1]

—Adm. Stansfield Turner, USN (Ret.)

CHAPTER 4

LESSONS FROM AIR AND SURFACE WARFARE HISTORY

There was also the inescapable truth that the Argentinian command-ers failed inexplicably to realize that if they had hit [the carrier] Hermes, *the British would have been finished. They never really came after the one target that would surely have given them victory.*[1]

—Adm. Sir John "Sandy" Woodward, RN,
Falklands British Task Force Commander

On the afternoon of April 6, 1945, the crew of the Japanese battleship *Yamato* assembled to sing the national anthem and give three banzai cheers for the emperor. The battleship, accompanied by the light cruiser *Yahagi* and eight destroyers, had just departed Tokuyama Bay to execute Operation Ten-Go. The mission called for the force, under the command of Vice Admiral Seiichi Ito, to attack the fleet supporting the American landings on Okinawa. Carrying only enough fuel for a one-way trip, the ships were to beach themselves to serve as shore batteries and to allow their crews to fight on as infantry.[2] The fleet's attack was also intended to distract the enemy's planes enough to allow Japanese bombers and kamikazes through the defenses.[3]

Unfortunately for Admiral Ito, the element of surprise would not be on his side. Adm. Raymond Spruance and the Fifth Fleet knew from communication intercepts the Japanese group's composition and plan, and the submarines USS *Threadfin* and USS *Hackleback* detected and reported the Japanese force just hours after their departure.[4] Once scout planes located the Japanese ships the following morning, Vice Adm. Marc Mitscher launched a massive air attack consisting of 386 U.S. planes.[5] In the course of two hours these veteran pilots conducted devastating attacks on *Yamato* and her escorts. The Japanese ships put up as much antiaircraft fire as they could, but without any fighter protection they could not prevent the American planes from precisely coordinating devastating bomb and torpedo runs. After being hit with at least a dozen bombs and torpedoes, the *Yamato* ammunition stores detonated as she slipped beneath the waves. Admiral Mitscher's pilots had sunk the mighty *Yamato*, along with *Yahagi* and three other destroyers, long before they were able to bring their guns to bear on the Allied ships. The Americans only lost ten planes in the process.[6]

The Japanese also launched massive air attacks during the weeks of fighting over Okinawa. One of the largest raids, dubbed Operation Kikusui I, was carried out the same day *Yamato* set sail. More than 400 planes, including 289 kamikazes, attacked the carriers, escorts, amphibious forces, and logistics ships. Despite hits on 19 ships and the destruction of 3 destroyers, the U.S. fighters and antiaircraft gunners protected the capital ships from serious damage, shooting down 135 Japanese planes in the process.[7] These attacks would continue for weeks to come and would score some successes, such as when two kamikazes hit the carrier USS *Bunker Hill*, killing 396 men.[8] However the fleet was able to sufficiently defend itself and retain sea control.

In addition to the surface and air attacks, the Japanese also dispatched a strong submarine force to disrupt the Okinawa landings. In one engagement, the destroyer USS *Stockton* detected the surfaced submarine *I-8* by radar on the night of March 31. When *I-8* submerged, *Stockton* and the destroyer USS *Morrison* used depth charges to force the Japanese submarine to the surface, where the destroyers sank it with gunfire. The U.S. Navy sank three more Japanese submarines that deployed to the area, as well as two of the four submarines carrying kaiten suicide mini submarines.[9] The fleet's antisubmarine escorts were able to form a strong ring around the fleet and allow the carriers to execute their mission. After almost three months of bitter fighting and tens of thousands of deaths, American forces captured the island in late June 1945.

The Battle of Okinawa was important to the present study for two key reasons. First, in a war filled with similar evidence, it provided final proof of the supremacy of the carrier task force. The aircraft carrier confirmed its dominance in all its primary mission areas, including reconnaissance, anti-surface warfare, antisubmarine warfare, air defense, and strike warfare. It won control of the sea by sinking the fearsome *Yamato*, shooting down hundreds of enemy bombers, and contributing to one of the deadliest months of the war for Japanese submarines. It used that sea control to project power ashore and enable U.S. Marine Corps and Army forces to capture the island. At the conclusion of the battle and the war, there was little doubt that the aircraft carrier was the most powerful ship type afloat.

Unbeknownst to its participants, the Battle of Okinawa was also important because it was the last time in naval history that an aircraft carrier has been forced to fight for sea control or defend itself against a credible attack. Since then, the U.S. Navy, led by its aircraft carriers, has played key roles in fights in Korea, Vietnam, Iraq, and Syria, and thousands of sailors have lost their lives in combat. However, the carrier has not been seriously challenged in mission areas such as open-ocean combat reconnaissance, anti-surface warfare, antisubmarine warfare, or integrated air and missile defense since the Battle of Okinawa, a span of more than seventy years.

During those seventy years of relatively peaceful seas, the U.S. Navy has used every available resource to design its force structure. It has used wargames, tabletop and at-sea exercises, technical reviews, and intelligence analyses to inform its decisions on how to design, build, and operate a fleet to deter and win the nation's wars. Despite these valuable inputs, the Navy for the most part has not been able to benefit from the most important resource in fleet design: the lessons of recent naval combat. Combat is the gold standard for which platforms, weapons, and tactics are effective and which have become obsolete. Unless recently tested in battle, any ship, weapon, or warfighting method is suspect, no matter how impressive they may appear in peacetime.

For example, during his presidency, Thomas Jefferson believed that a fleet of small coastal gunships would provide a much better defense than a powerful blue-water navy. He was able to sustain this conviction until the War of 1812, when combat proved his boats to be completely ineffective. Following the extensive trench warfare of World War I, many militaries believed large fixed defensive fortifications would dominate in subsequent wars. It was not until armies repeatedly bypassed or overcame these defensive systems, such as the Maginot Line, that combat proved that strategy wrong. The U.S. Navy's report on the 1982 Falklands Conflict, the one example of major naval combat since the Battle of Okinawa, highlights the value of learning from combat. It states that when considering modern threats there are multiple ways to prepare for them, including "exercises, war games, studies, intelligence collection, etc. But there is no substitute for actual combat experience."[10]

This chapter reviews recent "actual combat experience" to understand its lessons and how they should inform U.S. Navy force structure today. This review starts with the hypothesis that the aircraft carrier was the most powerful ship afloat in 1945, a hypothesis conclusively proven by extensive combat, including the Battle of Okinawa. From that starting point, the chapter examines the historical combat evidence to determine if the aircraft carrier remains the preeminent platform today. That analysis focuses on examples of fleets engaging in their primary wartime mission areas of reconnaissance, anti-surface warfare, integrated air and missile defense, and strike warfare, with antisubmarine warfare examined in Chapter 6. This review, examining both U.S. Navy and foreign navy combat, seeks evidence of the carrier's utility across all mission areas. It also seeks evidence of the carrier's ability to defend itself in combat.

That examination progresses to the present, when the aircraft carrier is purported to be the virtually indestructible, unquestionable flagship of the fleet. Yet, what concrete warfighting proof is there that the aircraft carrier is still the dominant ship class? Is the aircraft carrier king of the ocean because combat has proven it to be, or simply because it was in the last war and there have been no major battles to prove otherwise since? Is the carrier the centerpiece of the U.S. Navy because it is truly the best option, or because it is the option that naval officers, elected officials, and the public are most accustomed to?

These questions lead to uncertain, uncomfortable answers. In the last seventy years, there is little combat proof of a carrier's ability to win sea control or defend itself against modern threats. Though the carrier has delivered incredible firepower against land targets in numerous fights over that span, it has done so from seas uncontested since the Battle of Okinawa in 1945. The U.S. Navy's unclassified report on the Falklands Conflict acknowledges that since World War II, U.S. military operations have been "relatively immune to enemy interference."[11] Given this lack of combat evidence, it is impossible to conclusively say that the modern aircraft carrier is the ultimate ship and deserves to continue to lead the fleet into the future. On the other hand, because of that lack of recent combat experience, no one can conclusively say that the aircraft carrier is indisputably obsolete. However, the last seventy years of combat data does provide strong evidence of the aircraft carrier's waning usefulness and increasing vulnerability.

THE U.S. NAVY IN COMBAT

Since World War II, the Korean War offers the first case of potentially informative naval conflict. During the war U.S. Navy aircraft carriers launched large numbers of air strikes and provided invaluable support to the Pusan Perimeter early in the war before other sources of airpower were available. The U.S. Navy suffered losses to mines and land-based artillery but experienced almost no Korean attacks on its ships, easily controlling the seas throughout the conflict. In what is believed to be the only air attack on U.S. ships during the war, two Yak-3 bombers attacked the cruiser USS *Rochester*, hitting the ship's crane with a bomb that failed to explode.[12] Adm. James L. Holloway III, a carrier pilot during the war and later the Chief of Naval Operations, wrote, "Maritime forces operated with impunity off the coasts of Korea, launching air strikes, conducting shore bombardments, reinforcing troops, and delivering combat logistics, all in support of the [United Nations] forces ashore."[13] Because of that impunity, the war offers little proof of the carrier's ability to win sea control or defend against sustained attacks, although it did provide further evidence of the carrier's unmatched power projection capabilities in benign environments.

The blue-water navy played a somewhat similar role in the Vietnam War, again projecting power from uncontested seas. The fleet provided gunfire support, moved substantial numbers of troops and supplies, and launched aircraft that flew more than half the strike missions against targets in North Vietnam throughout the war.[14] However, when attempting to learn about a carrier or surface ship's ability to defeat enemy naval forces or defend against an attack, there are few meaningful lessons throughout the war. There was the famous Tonkin Gulf incident, in which three Vietnamese patrol craft unsuccessfully attacked the destroyer USS *Maddox* on August 2, 1964. Two days later, *Maddox* and the destroyer USS *Turner Joy* underwent what was most likely an imagined attack brought on by poor weather, nighttime conditions, and the ineffective use of sonar and radar aboard the destroyers.[15] Later in the war, on April 19, 1972, in Dong Hoi Gulf, North Vietnamese MiG-17s attacked four warships. During the short action, one of the attacking MiGs landed a bomb on one of the gun mounts on the destroyer USS *Higbee*. Fortunately, the mount had just been cleared due to a misfire, so no lives were lost. The frigate USS *Sterett* used a Terrier missile salvo to shoot down at least one MiG.[16] These small, short actions were some of the few times U.S. ships were attacked during the entire war, meaning the Vietnam War is a poor resource to study surface ships' reconnaissance, anti-surface warfare, antisubmarine warfare, and air defense capabilities.

In 1986 Libyan dictator Muammar Qaddafi's sponsorship of terrorist attacks and an illegal claim to the Gulf of Sidra prompted the United States to send a fleet to cross his declared "Line of Death."[17] The U.S. force, which included three aircraft carriers, twenty-seven escorts and support vessels, more than two hundred planes, and multiple submarines, quickly destroyed several key Libyan targets with relative ease.[18] In this case a brief, lopsided affair yielded little information in the way of understanding the fleet's capabilities in mission areas other than strike, as the vastly overmatched Libyans only had several patrol boats and airplanes that could mount a half-hearted attack.

The following year the frigate USS *Stark* was deployed in the Persian Gulf escorting merchants during the "Gulf Tanker War." On May 17, 1987, an Iraqi F-1 Mirage aircraft fired two Exocet missiles at *Stark*. Both missiles hit the ship, but only one exploded, killing thirty-seven sailors. The French-built AM-39 Exocet was a fearsome weapon. It attacked its targets using radar-homing while speeding at approximately Mach 0.93 just ten to fifteen feet above the ocean surface.[19]

Although this event is an isolated attack against a single ship, it illustrates several important points. Despite *Stark*'s various defensive systems, including SM-1 MR missiles, a Mark 75 76-mm gun, .50-caliber guns, the vaunted close-in weapon systems (CIWS), and Super Rapid Bloom Offboard Chaff (SRBOC), none were employed to prevent both missiles from hitting their target.[20] While these failures were due to a combination of poor watch-standing and the attack coming as a surprise, it is also important to note that no radar detected the incoming missiles and the officers and crew did not realize they were under attack until lookouts spotted the first missile inbound.[21] Thus, while these defensive systems may have protected the ship, the simple fact is that a modern U.S. frigate suffered a mission kill in part because it did not even detect the attack. Given the limited scope of the action and the extenuating circumstances, one should not draw too many conclusions from the attack on *Stark*. However, it is an important data point showing that even a single fighter, when armed with anti-ship missiles, can disable modern warships. Whereas the U.S. Navy needed 386 planes to destroy *Yamato* and her escorts in 1945, the Iraqis needed but a single aircraft to cripple *Stark* just forty-two years later, capitalizing on the benefits of the Missile Age.

Less than a year after the attack on *Stark*, the frigate USS *Samuel B. Roberts* struck a mine in the international waters of the Persian Gulf. The Navy determined other mines in the area to be Iranian, and President Ronald Reagan ordered "U.S. forces in the Persian Gulf to launch retaliatory strikes against Iranian military units in the Persian Gulf."[22] In what was dubbed Operation Praying Mantis, nine U.S. warships and a carrier air wing aimed to destroy Iranian oil platforms and any forces that attempted to defend them.

All taking place on April 18, 1988, the fighting started when two of the three surface action groups successfully destroyed two Iranian oil platforms. Soon after, the Iranians began launching uncoordinated yet bold individual attacks. The French-made attack boat *Joshan* launched a surface-to-surface Harpoon missile at the cruiser USS *Wainwright*. *Wainwright* narrowly avoided the missile through a combination of chaff, electronic countermeasures, and evasive maneuvers with the Harpoon passing a scant one hundred feet from the cruiser. The nearby frigate USS *Simpson* immediately fired SM-1MR Standard missiles at *Joshan*, with four of the missiles hitting the Iranian craft and causing it to sink later in the day.[23] Soon after *Joshan*'s failed attack, the Iranian frigate *Sahand* left port to conduct its own attack. *Sahand* took a severe beating from multiple attack aircraft as well as missiles and gunfire from the nearby surface action group before sinking later that night. The final attack came from *Sahand*'s sister ship *Sabalan*, which was devastated by planes from USS *Enterprise*.[24]

Operation Praying Mantis offers some lessons, but because it was so short and with such grossly mismatched forces, it is difficult to draw too many conclusions. Despite the captain of *Enterprise* recognizing the engagement as "the largest American sea battle since World War II," the fighting only lasted one day, making it a very small sample to study.[25] Furthermore, the Iranian force was severely outgunned and outnumbered and launched their attacks piecemeal. Even with these disadvantages and miscues the Iranians nearly managed to score a hit on *Wainwright* using just a single unsupported patrol craft.

The final major U.S. air-sea engagement of the last seventy years occurred during the Persian Gulf War. In the opening stage of the war, the Iraqi navy attempted to escape to Iran. In what became known as the "Bubiyan Turkey Shoot," coalition forces destroyed or damaged seven Iraqi missile boats, three amphibious ships, a minesweeper, and nine other vessels in the space of just five days.[26] This was another case of a short-lived, one-sided engagement where U.S. sea control was never tested. As Adm. William Owens, former vice chairman of the Joint Chiefs of Staff, wrote about Desert Storm: "No opposing naval forces challenged us. No waves of enemy aircraft ever attacked the carriers. No submarines threatened the flow of men and materiel across the oceans."[27] The engagement did feature the only time in history that a surface-to-air missile shot down an anti-ship cruise missile, when HMS *Gloucester* used a Sea Dart to intercept a Silkworm surface-to-surface missile fired at the battleship USS *Missouri*.[28]

Thus concludes the U.S. Navy's history of major conventional combat over the last seventy years. Over that span, the ships of the U.S. Navy have never had to fight for sea control and have rarely if ever faced a sustained, credible threat to their safety. The aircraft carrier has repeatedly proven its ability to project power ashore throughout that period, but its ability to execute almost all other warfare mission areas has gone untested. While peaceful seas are undeniably a good problem to have, the resulting dearth of naval conflict means there is little combat evidence that the aircraft carrier remains the dominant weapon it was in 1945.

FOREIGN NAVIES IN COMBAT

With so few battles in the U.S. Navy's modern history, one would naturally look to other nations' wars to attempt to fill the gaps. Yet here again, with one key exception, there are few useful data points.

Many foreign naval engagements featured small patrol craft and anti-ship missiles in the littorals. For example, following the Six Day War in 1967, an Egyptian Komar missile boat used Soviet Styx missiles to sink the Israeli destroyer INS *Eilat* as it sailed near Egypt's Port Said in what was the first wartime use of an anti-ship missile.[29] In the 1971 Indo-Pakistan War an Indian task force led by Osa-II class missile boats attacked the Pakistani naval base at Karachi using Styx missiles to sink a destroyer, a minesweeper, and two merchants.[30] In the 1973 Yom Kippur War, missiles again played a prominent role. In one engagement, Israeli Sa'ar missile boats successfully defeated Styx missiles fired by Egyptian missile boats, countering with Gabriel missiles that sank or heavily damaged three of the four Egyptian vessels.[31] Throughout the entire war, nineteen of fifty Gabriel missiles hit their targets, while the Israelis evaded all forty-seven of the Egyptian-launched Styx missiles.[32]

There was also wide use of anti-ship missiles in the Iran-Iraq wars of the 1980s, with 187 ships suffering a missile or rocket attack between 1984 and 1987.[33] In 2006, Hezbollah forces in Lebanon fired a Chinese-designed C-802 surface-to-surface missile at the Israeli corvette INS *Hanit*. The corvette suffered four deaths and extensive damage but was able to make it back to port for repairs.[34] In 2015 the Egyptian branch of the Islamic State launched a missile attack on an Egyptian Swiftships patrol boat. While many of the details are unclear, the militants are rumored to have used a Russian-made anti-tank guided missile fired from land.[35] Finally, in 2022 Ukrainian ground forces launched two Neptune anti-ship missiles which sank the Russian guided-missile cruiser *Moskva*, flagship of the Black Sea Fleet, equipped with multiple air defense systems.[36]

Each of these actions are worth study, but they offer limited lessons because they tended to involve small forces, unconventional attacks, and were short in duration. They did show that missiles can level the playing field, enabling small patrol craft to challenge destroyers and empowering weak irregular forces, such as Hezbollah or Houthi rebels operating from land, to attack more powerful conventional forces at sea. These fights also demonstrated that small missile craft executing "soft-kill" defenses, such as decoys and jamming, can evade large numbers of anti-ship missiles, as the Israelis did during the Yom Kippur War.[37] Finally, these actions showed that when defenses fail against missiles, the results can be devastating, with just one hit often resulting in a mission kill or sinking.

Those various battles offer some lessons, but their value pales in comparison to that of the 1982 Falklands Conflict. That clash, featuring sustained conventional combat between two near-peer adversaries, is the best available modern case study for examining fleet design, aircraft carrier utility, and surface ship survivability. It featured the first use of modern cruise missiles against a major navy, the only sustained air attacks against warships since World War II, the only use of nuclear submarines in combat, and the first use of vertical/short takeoff and landing (VSTOL) aircraft in combat.[38] In the U.S. Navy's search for combat evidence of the carrier's utility or vulnerability, the Falklands Conflict offers the most robust and modern data available.

The Falkland Islands, located approximately 650 km east of mainland Argentina, are small, bitterly cold, and rocky. The islands had been a British territory since 1833, yet in 1982 tensions were rapidly escalating between Argentina and Britain over their control.[39] Many Argentines believed the Malvinas—as the islands were known in Argentina—should become part of their nation due to their proximity to the mainland and the demise of colonialism around the world and within the British Empire. The Argentinian military junta that led the country pushed these claims further to unite the country and distract the populace from numerous domestic problems. When diplomatic negotiations failed, Argentina invaded the Falklands on April 2, 1982, easily capturing the lightly defended islands.[40]

The British immediately put plans into motion to fight for the islands. The Royal Navy dispatched three nuclear-powered submarines—*Spartan*, *Splendid*, and *Conqueror*—to make the long transit to the Falklands.[41] Once the surface warships completed refitting and resupplying, they too set sail for the South Atlantic. Those ships split into one fleet that would establish air and naval superiority, led by the small carriers HMS *Invincible* and HMS *Hermes*, and another that would land the amphibious forces to recapture the islands.

The British surface fleet planned to employ a layered defense against Argentinian surface and air attacks. In the first layer of defense the fleet had twenty Harriers, a short takeoff vertical landing (STOVL) aircraft that had not yet been tested in combat but proved to be a highly effective fighter. Farther in, the Sea Dart and Sea Wolf missile defense systems, installed on the modern Type 42 and Type 22 destroyers, respectively, provided the next layer of defense. Finally, close-in systems and soft-kill measures formed the final defensive ring. The escorts had an array of antiaircraft guns such as the 4.5-inch gun and the 20-mm Oerlikons, and many ships employed a chaff system to decoy incoming missiles.[42] Thus, while sailing far from home the British fleet had an impressive, layered air defense led by the new Type 42 destroyers, although the fleet suffered from the lack of airborne early-warning aircraft.[43] All five of the Type 42s were within four years of their commissioning.

Awaiting this force was an Argentinian air force and navy, both of which were capable but relatively small and equipped with old equipment. The majority of the thirteen major surface combatants were outdated, although they had two of the same Type 42 destroyers, three French A69 frigates, an aircraft carrier, and an old cruiser (ARA *General Belgrano*). The Argentinians also possessed several submarines, including two German-built Type 209 diesel boats.[44] In the air the Argentinians had approximately a hundred operational aircraft, including fourteen French-built Super Etendards capable of launching Exocets.[45] At the start of the hostilities the Argentinians only possessed five air-launchable Exocets.[46]

After recapturing the island of South Georgia, the British set up around the Falklands and started to work on establishing air and naval superiority. The Harriers made several attacks on installations and the airfield on the Falklands. Simultaneously, destroyers and frigates bombarded positions around Port Stanley, the capital of the Falkland Islands. Both sides suffered casualties in the air as they tested each other's defenses and explored the capabilities of their own weapons.

On May 2, just days after hostilities began around the Falklands, the submarine HMS *Conqueror* found and attacked the cruiser ARA *General*

Belgrano, sinking the ship and killing 368 Argentinians.[47] This attack proved to be pivotal. Beforehand, casualties had been minimal as each side had slowly ramped up the fighting, but the sinking of *Belgrano* made it clear to the world that the diplomatic stage was over and a war had begun. In addition, by sinking the cruiser, the British neutralized a powerful threat to their surface fleet and carriers. Most importantly the sinking convinced the Argentinian navy to withdraw its ships back to port for fear of additional submarine attacks. This ensured that the remainder of the war was primarily fought between British air and naval forces against only Argentinian air threats.

Despite the Royal Navy's initial success, Argentina evened the score just the next day. Finding British escorts approximately 110 km southeast of Port Stanley, two Super Etendard fighters made a low-level approach, firing one Exocet at the destroyer HMS *Sheffield* and one at nearby HMS *Yarmouth*.[48] Only two and a half minutes after the initial radar detection, one of the Exocets slammed into *Sheffield*, igniting fires, knocking out the fire main, and killing twenty sailors. The officers and crew were eventually evacuated and after several days of continued burning and a worsening list, *Sheffield* sank. Journalists Max Hastings and Simon Jenkins captured the British reaction well when they wrote, "Officers and men alike were appalled, shocked, subdued by the ease with which a single enemy aircraft firing a cheap—£300,000—by no means ultra-modern sea-skimming missile had destroyed a British warship specifically designed and tasked for air defense."[49] The *Sheffield* sinking was an incredibly important event. A modern British warship (only three years old) with a respectable air defense system, in a heightened defensive posture and expecting attacks from Argentinian planes and missiles, had been sunk by one missile fired by one aircraft with the entire action taking place in less than three minutes.

Despite the loss of *Sheffield*, the British task force continued its efforts to gain control of the area around the Falklands. Over the next several weeks, British Harriers and missile defense systems shot down seven aircraft, while the Argentinians inflicted considerable damage on the Type 22 frigate HMS *Brilliant*, forcing it to withdraw for the remainder of the war.[50]

On May 21 the British commenced an amphibious landing at San Carlos, on the western side of East Falkland Island. Despite not having gained air superiority, the British war cabinet decided to go forward with the attack because of the impending harsh South Atlantic winter and mounting political pressure. The approaches to the beach were confined, making the transports and their escorts vulnerable to air attack while unloading troops and supplies. During this stage, despite the Harrier combat air patrol, the Sea Wolf and Sea Dart interceptors, and various other antiair defenses, British ships took heavy losses.

The first victim was the Type 21 frigate HMS *Ardent*, struck by two 1,000-pound bombs on May 21. Two days later another Type 21 frigate, HMS *Antelope*, was attacked by Skyhawks and sank soon after. On May 25, the Type 42 destroyers HMS *Coventry* and HMS *Broadsword* were operating farther out to sea on picket duty. As Skyhawks closed in for the attack, Sea Wolf missiles from *Broadsword* failed and a Sea Dart fired from *Coventry* missed. Multiple 1,000-pound bombs mutilated *Coventry*, which sank soon after.[51] On the same day, SS *Atlantic Conveyor* was sailing in open waters approximately 110 km northeast of the islands under the protection of the carrier escorts. *Atlantic Conveyor* was a merchant ship transporting much-needed Harriers, helicopters, and other supplies to the fleet. In an attack similar to that on *Sheffield*, two Etendards fired one Exocet each. HMS *Ambuscade* launched its chaff defenses, likely causing the missiles to veer away from the Type 21 frigate but toward the vulnerable *Atlantic Conveyor*, hitting the large target and causing her to sink several days later.[52] After that crucial day Argentinian attacks tapered off, although they still managed to sink the landing ship RFA *Sir Galahad*, and inflict significant damage on the destroyer HMS *Glamorgan* and landing ship RFA *Sir Tristram*.[53] The remainder of the conflict focused on the land battle in which the Royal Marines spread out from San Carlos and eventually recaptured Port Stanley to put an end to hostilities.

The Falklands Conflict's lessons are wide ranging. In the reconnaissance mission area, the war demonstrated how difficult it is to hide large surface ships from a capable enemy. In the air, Argentine Boeing 707 transport—not reconnaissance—aircraft detected and tracked the British task force as it transited south, first gaining contact days before the British thought possible.[54] Super Etendard pilots analyzed Harrier radar contacts to surmise the location of the strike group, then used that data to launch the Exocet attack that destroyed the transport SS *Atlantic Conveyor*.[55] At sea, Argentina discretely employed five surveillance trawlers to report the position of the British task force.[56]

Today, potential U.S. adversaries' ability to detect and track a U.S. surface group are much greater than those of Argentina during the Falklands Conflict. For example, China has dozens more surveillance aircraft, a first-rate unmanned aerial vehicle program, and a robust satellite network that—according to the U.S. Office of Naval Intelligence—allows China to "observe maritime activity anywhere on the earth."[57] Additionally, China can rely on its massive People's Armed Forces Maritime Militia—hundreds of trawlers and merchants, camouflaged among the thousands of civilian ships in the Chinese near seas—to discreetly and accurately report U.S. warship locations.[58] If the U.S. Navy and Imperial Japanese Navy could use 1940s technology to find and sink thirty-one of each other's carriers in World War II, and if the Argentinians could find and track the British task force to sink four warships and two auxiliaries in the Falklands Conflict, then surely modern adversaries like China can find U.S. carrier strike groups using all the same tools, as well as satellites, advanced radars, huge maritime militia reporting networks, and cyber penetration tools. The aircraft carrier's best defense—using its mobility to prevent the enemy from tracking it and launching an attack—is a less viable option today than ever in its history.

The Falklands Conflict also offers important lessons in the mission areas of anti-surface warfare and integrated air and missile defense. There was little fleet-on-fleet action because the Argentinians withdrew their fleet to port after losing *Belgrano*, but there were numerous air and missile attacks on the British surface fleet.

First, anti-ship missiles proved highly effective in their first use against a major navy. Of the just seven Exocets the Argentinians launched in the entire war, three scored hits. Those three hits resulted in the destruction of two ships and significant damage to a third.[59] Whereas most Argentinian air attacks featured pilots brazenly flying above the wavetops to deliver unguided bombs against fiercely defended warships, Super Etendard pilots could safely launch their Exocet miles away from their targets with devastating results. The Falklands Conflict truly ushered in the Age of the Missile at sea, representing both a huge opportunity and liability for the U.S. Navy today.

Against those missiles, hard-kill defenses proved ineffective. The modern Sea Dart and Sea Wolf systems suffered multiple flaws and misses that enabled successful Argentinian attacks, and neither system (nor short-range weapons) intercepted any Exocets. Fortunately for the British, the French halted sales of Exocet missiles to Argentina at the start of hostilities.[60] Today it is unlikely a credible adversary will be so limited in quantity of offensive weapons; for example, a single Chinese Houbei fast-attack craft has more anti-ship missiles than the entire Argentinian military fired in 1982—and China is estimated to have thousands overall.[61] On the defensive side, the Falklands Conflict was consistent with much of the rest of modern naval history, indicating that soft-kill measures are more effective than hard-kill defenses, that neither are credible options against a concerted attack, and that it is critical to attack effectively first to avoid relying on porous defenses.

British casualties—four warships lost (and one a mission kill) plus two auxiliaries destroyed—indicate that when surface warships are hit today, modern munitions will typically incapacitate or sink them. Capt. Wayne Hughes examined this modern trend at length in *Fleet Tactics and Coastal Combat*, writing that most warships today have "little staying power" and that one or two hits by missiles will "put most warships out of action."[62] In the Falklands Conflict, multiple ship losses occurred even with Exocets and bombs failing to detonate due to incorrect fuse settings. British task force commander Sir John "Sandy" Woodward later wrote, "We lost *Sheffield, Coventry, Ardent, Antelope, Atlantic Conveyor,* and *Sir Galahad.* If the Args' bombs had been properly fused for low-level air raids we would surely have lost *Antrim, Plymouth, Argonaut, Broadsword* and *Glasgow.* And we were very lucky indeed that *Glamorgan* and *Brilliant* were still floating in mid-June."[63] The era of one-hit ships suggests that the U.S. Navy should shift away from capital ships to smaller ships. If almost all surface ships will sink or suffer a mission kill after being hit—meaning large and small ship survivability is roughly the same—a fleet of large numbers of small ships is more resilient than one of just a few capital ships.

The Falklands Conflict also showed the benefits and risks of using aircraft carriers in contested environments. On the positive side, the carriers *Invincible* and *Hermes* provided vital air cover that enabled the entire operation. The Harriers flew combat air patrol for the fleet and the forces ashore and downed thirty-one Argentinian planes by the end of the war.[64] Furthermore, they deterred numerous Argentinian air attacks, preventing easy strikes on ships that could have resulted in additional losses. Carrier aircraft were especially important because they were the only reliable source of British air power; the Royal Air Force's only land-based contribution was seven Vulcan bomber attacks flown from faraway Ascension Island, which required seventeen in-flight refuelings and had "virtually no impact."[65] The Falklands Conflict confirmed that the aircraft carrier excels at fleet air defense, a mission it can execute better than any other platform.

However, the war also showed the risks and opportunity costs of using that airpower for missions other than fleet air defense, such as strike against ground targets. The British lost five Harriers to enemy action throughout the war, all of them shot down by land-based antiaircraft defenses.[66] That was a high price to pay when Admiral Woodward only had approximately thirty-four aircraft available, all desperately needed for air defense.[67] After losing one of the Harriers in a strike against the airstrip at Goose Green, Admiral Woodward decided he "should not risk any more of our precious Harriers by allowing them to go out on these high-speed, low-level, cluster-bomb attacks against heavily defended Argentine positions. Quite simply, I could not afford to lose my strictly limited force of air defence aircraft . . . on this not-very-effective sort of task."[68] For the U.S. Navy today, this suggests that it may be best served by focusing its carrier airpower on fleet air defense—a mission only it can do—while shifting strike missions to other portions of the fleet, a role most platforms with missiles can better accomplish without risking pilots or their valuable aircraft against formidable air defense networks.

Finally, the Falklands Conflict demonstrated the risk of centralizing so much of a fleet's firepower on a small number of ships. The carriers *Hermes* and *Invincible* were the greatest assets in the British task force. They were also the greatest British liability, and they dictated the deployment and tactics of the entire task force. Woodward wrote of the "inescapable truth that the Argentine commanders failed inexplicably to realize that if they had hit *Hermes*, the British would have been finished. They never really went after the one target that would surely have given them victory."[69] Woodward's solution was to keep the carriers as far out to sea as possible, almost exclusively using them for air defense. The U.S. Navy now has the option to distribute much of the carrier's missions to other platforms, an alternative unavailable to Admiral Woodward. Alleviating the burden on the carrier would allow the use of smaller carriers, enabling them to operate farther out to sea to minimize the threat of enemy attacks, and reducing the impact if one is lost.

The Falklands Conflict, with fighting that lasted a brief three months and all confined to a single region, offers a small dataset of combat lessons compared to conflicts like the American Civil War, World War I, and World War II. However, because it was fought more recently than those conflicts and featured weapons and platforms not used extensively in combat before—such as missiles, STOVL aircraft, and nuclear-powered submarines—the Falklands Conflict is a "gold mine of lessons," as a former commander of the U.S. Atlantic Fleet put it.[70] Some of its most important lessons are focused on the missile and the aircraft carrier. Anti-ship missiles proved lethal in their first major use in combat, with a small number of them having a large impact on the fighting. Their performance supports the development of fleet structures like the Flex Fleet, which maximizes the use of missiles and the number of missile-launching platforms, shifts resources to soft-kill measures over interceptors, and focuses on attacking effectively first as the most effective defense against missiles. The Falklands Conflict also showed how aircraft carriers are still important parts of the fleet, but that they do not have to accomplish all the fleet's missions. By focusing the carrier on fleet air defense and shifting many of its other missions to different platforms, each ship in the fleet can execute a mission it does best. That mission reduction means the carriers in structures like the Flex Fleet can be smaller, less expensive, more numerous, and thus more effective and survivable as a group.

LESSONS FROM RECENT NAVAL COMBAT

This book questions the carrier's status as the flagship of the U.S. Navy. Part of that examination looks forward from present day, seeking to understand the fleet structures that can replace the carrier-centric model. That review, in Part I, focuses on the opportunities afforded the U.S. Navy if it reduces its reliance on the large nuclear-powered aircraft carrier. The second part of that examination looks from present day to the past, seeking evidence of the carrier's utility or vulnerability. That review, in Part II, seeks hard proof of the carrier's ability to execute its mission and defend itself. In that analysis, naval combat is the most important type of evidence.

Yet with little naval combat to study and learn from since World War II, it is exceedingly difficult to know what fleet designs and ship types are best suited for the next war. Early in the U.S. Navy's history there were somewhat frequent naval conflicts to learn from: the American Revolution, Napoleonic Wars, Quasi War, Barbary Wars, War of 1812, Crimean War, American Civil War, Spanish-American War, Russo-Japanese War, World War I, and World War II. However, since 1945 the Falklands Conflict is only major conventional war to study, and even that engagement was only three months long. This means that throughout the United States' existence, there was a major naval war occurring somewhere in the world in approximately 28 percent of the years before 1945, but only during roughly 1 percent of all subsequent years. As a result, the U.S. Navy's force structure decisions must be based on much less robust data than combat experience. In an award-winning U.S. Naval Institute *Proceedings* article, Barrett Tillman described these seven decades of peace as the "Post-Naval Era." He summed up the problem facing today's admirals well: "Today, as then, a new generation of ships, weapons, tactics and doctrine remain untried in combat. In fact, entire generations of naval hardware have come and gone with not one drop of blood shed. We never know how well we have prepared until we have to sink and perhaps be sunk."[71]

Power projection missions are the exception to this rule. Throughout this era, the U.S. Navy has struck targets in places like Korea, Vietnam, Libya, Serbia, Iraq, and Afghanistan, delivering firepower to support forces ashore. Yet in all those circumstances the fleet was able to launch attacks from its sanctuary at sea, facing virtually no threat until its pilots encountered land-based defenses and fighters. Given the frequency of these power projection missions, it is no surprise that designers of the *Ford* class of carriers sought to maximize its SGR.[72] Yet that metric only optimizes the carrier's ability to conduct one of its many mission areas, and only under ideal conditions.[73] As Capt. Robert C. Rubel, a former dean of naval warfare studies and professor at the Naval War College, wrote, "From Korea to Operation Iraqi Freedom, power projection operations conducted in benign sea control environments—usually by individual carrier strike groups—have been touted as warfighting."[74] If the U.S. Navy only expects to execute peacetime presence operations and benign-area strike missions in the future, then the carrier-centric fleet is the best option. However, if it wants to prepare for high-end combat, the U.S. Navy needs to evolve its force structure.

Other than frequent power projection operations, the aircraft carrier has not executed its primary warfighting missions since the Battle of Okinawa. No carrier has completed extensive combat reconnaissance, anti-surface warfare, antisubmarine warfare, or integrated air and missile defense. No carrier has faced platforms and weapons that are commonplace today, such as anti-ship missiles, nuclear-powered submarines, jet aircraft, and satellite networks. As Captain Rubel also wrote, "The danger is that the net outcome of an offense-versus-defense battle cannot be truly known short of actual fighting. Thus, the ability of aircraft carriers, as capital ships, to carry out the command-of-the-sea mission is increasingly being placed in question."[75]

What little fighting has occurred suggests the carrier's utility and survivability is waning. A key part of the carrier's defense is its ability to stay mobile to avoid targeting by the enemy, but today it is harder to stay hidden at sea than at any other point in naval history. Once the carrier is found, its escorts need to provide near-perfect protection for such a valuable asset, but campaigns such as the Falklands Conflict and the attack on USS *Stark* show how difficult it is to defend against anti-ship missiles. When those missiles hit their target, recent history demonstrates how lethal modern munitions can be, with even large ships likely suffering a mission kill if they manage to avoid sinking. Many of these trends are not new in naval history; it has always been difficult for ships to remain hidden at sea, defenses have always been temporary, and ships have always been susceptible to sinking. The difference today is that most of the U.S. Navy's firepower is concentrated in just ten nuclear-powered aircraft carriers, meaning their defenses must be perfect. There is little evidence in modern history—or any naval history—that the U.S. Navy can realistically develop such a defense.

In a famous *Naval War College Review* article, Rear Admiral Yedidia "Didi" Ya'ari discussed the lack of modern naval combat available for study: "An important fact to keep in mind about sea power is that it has not been truly tested since World War II." Describing the consequences of that fact, he wrote,

> There exists, then, the quite peculiar situation in which nothing can be properly substantiated—neither commonsense-based adherence to the experience and convictions of sea power that we have long had, nor the intuitive feeling that the cumulative change is now of such magnitude that a radical rethinking of these convictions is of crucial importance. Yet the latter is a matter of more than intuition, at least in one sense—we are entering the next century, and also a dramatic transition from the open ocean to the shallow seas, with a severe lack of relevant experience. Too often in military history, at such moments of uncertainty old principles have hardened into beliefs that have been, among other things, the grounds for rejecting new, more adequate ones. We have no option but to rely on an analytical process that subjects every conviction of the past, including the most fundamental ones, to unconditional examination.[76]

This book is an attempt to use "an analytical process," relying on data from history and current events, to subject the aircraft carrier to "unconditional examination." One result of that examination is the Flex Fleet, an example of an improved fleet structure that could help the U.S. Navy retain its dominance. The Flex Fleet is built with the historical lessons of this chapter in mind. Learning from the historical successes of Exocets and other anti-ship missiles, the Flex Fleet is built to maximize the use of these weapons on large numbers of small, yet powerful, ships. The Flex Fleet calls for fewer large ships such as *Ford*-class carriers and *Arleigh Burke*–class destroyers because of small ships' increased offensive power, the difficulty associated with hard-kill missile defenses, and because of all surface ships' poor staying power. The Flex Fleet does not rely on aircraft carriers for all combat operations, but instead uses them primarily for fleet air defense—just as the British did in the Falklands Conflict.

Following the Battle of Okinawa and World War II, the world's navies had accrued years of combat experience and so could confidently know what did and did not work in battle. Since then, as former Naval War College president, Rear Adm. Walter Carter Jr., wrote, "More than 70 years have passed since a major maritime conflict. During that time many new technologies have emerged, with few combat tests to provide reliable guideposts regarding what will prove successful now and in a future conflict."[77] What "guideposts" are available indicate the aircraft carrier, unquestionably dominant at the Battle of Okinawa, is an increasingly questionable flagship for the U.S. Navy today.

ᘒCHAPTER **5**

AIR AND SURFACE WARFARE TODAY

The surface ships now in commission were designed with the open ocean and distant defensive perimeters in mind; to keep deploying them to a playing field where, under the most optimistic assumptions, their survival requires as a normal operating mode the highest level of everything, all the time, is unhealthy and unrealistic in the long run.[1]

—Adm. Yedidia Ya'ari, Israeli navy

Today, a formidable China and resurgent Russia are forcing the United States to question its assumptions about the future of naval warfare. The U.S. Navy can no longer assume that its carriers can operate from a sanctuary at sea, nor can it assume that presence missions and power projection will remain its primary missions. The People's Republic of China (PRC) now has the world's largest navy, largest missile force, and third largest air force.[2] Meanwhile, Russia is fielding increasingly advanced platforms, weapons, and technologies, including Kalibr cruise missiles, *Severodvinsk*-class submarines, and world-leading cyber penetration tools.[3] If those militaries now have the potential to either destroy or deter the U.S. Navy's aircraft carriers, then the United States needs to either adapt how it fights or accept that it can no longer establish sea control when and where it desires.

This chapter seeks to assess the carrier strike group's ability to defend itself against peer adversaries, thus providing insight into the carrier's utility and future as the flagship of the U.S. Navy. That analysis uses the "archer-arrow" analogy common to naval defenses to determine the likelihood of U.S. carrier strike groups effectively defending themselves against air- and surface-launched attacks. Divided into stages, this analysis includes the carrier strike group's ability to stay mobile and hidden from the enemy (hiding from the archer), the strike group's ability to attack first and prevent the enemy from launching weapons (shooting the archer), and finally, the ability of the carrier strike group to intercept or defend against launched ASCMs and anti-ship ballistic missiles (ASBMs) (shooting the arrow). That analysis uses the PRC as the "archer," as a proxy for any potential adversary.

That analysis concludes that it will be exceptionally difficult to defend the aircraft carrier against a peer competitor, and especially not to the standard needed for the centerpiece of U.S. naval operations. Today, an adversary does not need to defeat the U.S. Navy's hundreds of ships, submarines, and aircraft; it only needs to stop approximately ten aircraft carriers, and there are now a lot of ways to do that. While the carrier is the best-defended surface ship in the world, that defense is still not good enough for a ship that must survive and operate as intended for the U.S. Navy to prevail.

THE ARCHER

For much of its long history China was a land power with a weak or nonexistent navy, a trend that continued through the 1980s. However, two key events in the 1990s convinced Chinese leaders that they would have to make drastic changes and modernize their weak navy and air force if their nation was going to truly be a player on the world stage. The first came in the 1991 Persian Gulf War, in which coalition cruise missiles and jets with precision munitions, all coordinated by an impressive command and control structure, surgically destroyed the sizable Iraqi military.[4] The PRC recognized the threat such a modern force posed and saw that their massive military could face the same fate as the Iraqis. The second lesson prompting Chinese modernization came in 1996. The PRC attempted to deter Taiwanese moves toward independence by conducting missile tests and military exercises, and in response, President Clinton sent carrier strike groups to the region. The PLA, with no ability to contest the U.S. Navy's actions, realized that it would need to develop a military capable of defending at least the near seas to prevent U.S. ships from acting with impunity in the area.[5] Attempting to ensure that the U.S. Navy could not influence any future Chinese interactions with Taiwan, PRC leaders reportedly issued orders to develop a solution to the "carrier problem."[6]

Since then, the PRC has developed a vastly improved People's Liberation Army Navy (PLAN) and People's Liberation Army Air Force (PLAAF). According to the Department of Defense's 2020 annual report to Congress titled "Military and Security Developments Involving the People's Republic of China," the PRC has sustained twenty years of annual defense spending increases, a feat that has made it the second-largest military spender in the world.[7] RAND Corporation analysts determined that in 2003, only about 14 percent of Chinese destroyers and 24 percent of its frigates could be considered "modern." Those figures that have jumped to 65 percent and 69 percent respectively as of 2015, a figure that has grown even more since.[8] The PRC has been aggressively using that empowered military, regularly deploying ships worldwide, executing complex exercises, and supporting the country's expansive territorial claims.[9] In early 2013 Capt. James Fanell, then the U.S. Pacific Fleet's director of intelligence and information operations, summed up Chinese development in a controversial speech saying, "The People's Republic of China's presence in the southern China sea prior to 1988 was nearly zero. Now, in 2013, they literally dominate it."[10]

The PLAN's improvements are numerous, but its advances in ASUW are likely the most impressive. As of 2020 the PLAN has the largest fleet in the world with approximately 350 battle-force platforms, including two aircraft carriers, 32 destroyers, 49 frigates, 49 corvettes, and 86 missile patrol craft, as well as various amphibious ships and auxiliaries.[11] Retired Cdr. Bryan McGrath and analyst Timothy Walton, writing for the Naval War College, said, "Relative to the U.S. Navy, ASUW is likely the area where the PLAN is comparatively strongest. It has developed potent, distributed potential striking power by fielding over-the-horizon-targeting systems and outfitting its range of small and large surface combatants with advanced, longer-range supersonic and subsonic anti-ship cruise missiles (ASCMs) having improved electronic systems, multiaxial attack coordination, and terminal evasion maneuvers."[12] While the Chinese employ a variety of classes, their fleet is led by the Renhai-class cruiser, Luyang-class destroyer, Jiangkai-class frigate,

Jiangdao-class corvette, and Houbei-class missile craft. Each represents a culmination of various other designs, incorporating a wide array of domestic and foreign technologies, and they are being produced in large numbers.

As is true of the Chinese surface ships, the PLAN and PLAAF air fleets are rapidly improving ASUW threats. As of 2020 the PRC has a total of 2,500 total aircraft, 2,000 of which are operational combat aircraft, giving it the largest air force in Asia and the third largest in the world. Of these aircraft, the U.S. Department of Defense considers approximately 800 to be modern fourth-generation platforms.[13] Additionally, the PRC has a massive UAV fleet, including armed reconnaissance, low-observable, and transport variants.[14] RAND Corporation senior policy analyst David Shlapak, writing for the U.S. Naval War College, described China's changes between 1990 and 2010:

> The PLAAF's fighter fleet has undergone a remarkable modernization. In 2000, only 2 percent of PLAAF fighters were "modern," "fourth genera-tion" aircraft, comparable to the American F-15 and its contemporaries. Little more than a decade later, almost one in three of China's fighters can be considered modern. In fact, only the United States and Russia own more fourth-generation fighters than does the PLAAF; China has more modern fighters than Britain and France combined. Also, in 2011 China joined the United States and Russia as the only countries to fly a stealth aircraft, with the J-20.[15]

The Department of Defense, reviewing the PRC's modernization efforts, concluded that the PLAAF is "rapidly catching up to Western air forces."[16]

A common theme of the PRC's various surface warships and attack aircraft is the ability to employ advanced ASCMs. In assessing the lessons learned from the ASCM attacks on the Israeli destroyer INS *Eilat*, the British destroyer HMS *Sheffield*, the frigate USS *Stark*, and the Israeli corvette INS *Hanit*, the PLAN has made great strides in the international procurement, indigenous development and production, and deployment of missiles on a wide variety of platforms.[17] Naval War College professors Toshi Yoshihara and James R. Holmes described the allure of missiles for the Chinese in their book *Red Star over the Pacific*, writing, "The conviction that a single missile—a modern-day incarnation of the sling and stone David used to defeat Goliath—can inflict outsized damage on large, expensive platforms such as aircraft carriers adds force to the Chinese penchant for asymmetric strategies and tactics. . . . In recent years, some Chinese analysts have taken to depicting guided missiles as an 'assassin's mace,' a term similar to the Western concept of the 'silver bullet.'"[18]

The PLA has numerous modern missile systems installed in virtually all of its surface combatant platforms, many submarine classes, as well as the vast majority of its maritime strike aircraft.[19] Altogether, the PLA has dozens of different ASCM models and a total inventory estimated to number in the several thousands.[20] Compounding the missile problem, many analysts believe the PLA's ASCM arsenal has surpassed that of the U.S. Navy. The primary U.S. ASCM, the Harpoon, was first introduced in 1977 and has a range of only 124 km.[21]

Supplementing these ASUW platforms and weapons are ASBMs, such as the Dong-Feng 21D (DF-21D) and DF-26B. ASBMs launch from mobile ground units, travel outside the atmosphere, and then attack their targets at speeds and angles that would make it extremely difficult to defend against. They are based off of older medium-range ballistic missiles intended for ground attack, and the Department of Defense has reported the DF-21D and DF-26B to have ranges of 1,500 km and 4,000 km respectively.[22] Many of the technical specifications are unknown or must be inferred from Chinese research literature, but it is believed to use either an active-radar, infrared (IR), or laser-homing system, and the warhead is likely high-explosive, armor-penetrator, cluster, or electromagnetic pulse (EMP).[23] Regardless of the exact warhead type, the damage would likely be devastating. Impacting its target at supersonic to hypersonic speeds at steep angles, it would almost certainly put any large ship out of action or more likely sink it. In an era of "one-hit ships," where destroyers like HMS *Sheffield* sank when hit by one small Exocet which did not even explode, the chances of any major warship being hit by an ASBM and maintaining its combat readiness is slim.

To achieve that hit, however, the ASBM must perfectly execute an extensive kill chain, leading to it being dubbed a "system of systems." While some of the ASBM kill chain relies on robust, decades-old technology—such as mobile launcher units—other portions have never been successfully fielded before (this would include homing at hypersonic speeds during atmosphere reentry).[24] The entire weapon system relies on extensive targeting, command and control, communications, and targeting systems that must all operate together.

Despite the immense technological obstacles, the evidence indicates that the PRC has developed an operational ASBM. Dr. Andrew S. Erickson, a Naval War College professor and expert on the Chinese military, used the very first sentence in his book on the ASBM to state, "China's DF-21D anti-ship ballistic missile (ASBM) is no longer an aspiration. Beijing has successfully developed, tested and deployed the world's first weapons system capable of targeting a moving aircraft carrier strike group (CSG) from long-range, land-based mobile launchers."[25] In 2020 the U.S. Department of Defense plainly stated that the DF-21D gives the PLA "the capability to conduct long-range precision strikes against ships, including aircraft carriers, out to the Western Pacific from mainland China."[26]

An operational ASBM will have large implications for the conduct of naval warfare and the balance of power in the western Pacific. Professors Yoshihara and Holmes explored the implications of an operational ASBM saying, "In short, the aircraft carrier and the forward base—the two iconic symbols of U.S. preeminence in international politics, and two of the pillars of Mahanian sea power—could face obsolescence, much as the battleship met its demise with the advent of naval aviation. Small wonder the ASBM has generated such angst in Washington."[27] In 2005 retired Rear Adm. Eric McVadon summarized the impact of ASBMs, saying they would represent "the strategic equivalent of China's acquiring nuclear weapons in 1964."[28]

With "archers" like the PLAN, the U.S. Navy will no longer enjoy the uncontested sea control its aircraft carriers have benefited from for so long. The PRC's numerous advanced surface warships, maritime strike aircraft, sea-skimming ASCMs, and ASBMs are all highly capable, and as Captain Fanell declared, "Make no mistake, the PLA Navy is focused on war at sea and about sinking an opposing fleet."[29]

HIDING FROM THE ARCHER

Capt. Wayne Hughes' *Fleet Tactics* states, "At sea better scouting—more than maneuver, as much as weapon range, and oftentimes as much as anything else—has determined who would attack not merely effectively, but would attack decisively first."[30] Like all other ships, the aircraft carrier's ability to evade enemy scouting efforts is pivotal to its success in battle. The ship's best defense has always been its use of mobility to prevent enemies from tracking it long enough to launch an attack.

To evaluate the modern carrier's survivability and utility in combat, it is crucial to understand its ability to remain hidden at sea. If enemy forces cannot find the aircraft carrier, then it can likely execute its missions and continue to lead the U.S. Navy into battle. On the other hand, if the enemy *can* effectively track carrier strike groups, their survivability—and thus utility—would be greatly reduced.

To predict adversaries' ability to find and track the aircraft carrier today, a review of the trends in scouting technologies and methods is informative. Navies have proven successful at finding and attacking enemy ships throughout all naval history. Do today's tools make it easier or harder to conduct scouting at sea? Have ships themselves become more or less susceptible to scouting, due to changes in their design, operations, or mobility?

To answer these questions, each of the primary methods of ship detection, identification, and tracking is analyzed below. That analysis compares three time periods—the Age of Sail, World War II, and the present day—to gain a broad understanding of scouting trends over time. Those scouting methods include visual, radar, acoustic, human intelligence, electronic warfare, signals intelligence, and satellites. For each method, a warship's ability to evade detection throughout time is also evaluated.

In simplest terms, navies have been successfully finding and targeting enemy ships for thousands of years. Today, navies have all the same tools as well as incredible new scouting and communications systems, all searching for ships that are larger and more recognizable than ever before. There is little evidence that the aircraft carrier could overcome these trends in history and technology to stay hidden for any appreciable amount of time.

Visual

The first and oldest scouting method is visual searching. Ships of the line and frigates kept alert lookouts in the crow's nest to spot masts over the horizon, and in World War II picket submarines and destroyers were sent out as the eyes of the fleet to provide early warning of the enemy. Scout planes were even more dominant in that period. In the vast expanse of the Pacific theater, Japanese and American forces sent thousands of carrier- and land-based aircraft to search the seas looking for the enemy. Today, large numbers of long-range aircraft, ranging from maritime patrol aircraft to UAVs, will scour the sea for targets like carrier strike groups.

Once these platforms locate the carrier, modern networks solve many of the problems that hampered scouting in the past, including poor communications and inaccurate ship identification. Instead of a single pilot using the naked eye to identify and report a fleet while under fire, aircraft and UAVs can immediately transmit pictures and videos to analysts ashore to accurately classify surveilled ships. As for the target, warships have only gotten larger and more recognizable as time has passed. In the Age of Sail a frigate could conceivably be confused for a merchant ship. Today, even an untrained eye could likely identify a 100,000-ton aircraft carrier or one of its escorts with little difficulty.

A retired four-star admiral and former commander of U.S Pacific Command disagreed with this analysis in a 2015 U.S. Naval Institute *Proceedings* article. The admiral argued that an aircraft carrier could be disguised as a supertanker or large container ship by using "imaginative lighting and some other deceptive tactics."[31] However, most warships and planes have infrared and night vision sensors that would quickly uncover the carrier's attempt at deception. Supertankers do not steam in formation with five other ships, and they do not carry out frequent course and speed changes common to carrier strike groups. Supertankers have radically different acoustic and electronic signals signatures than those of aircraft carriers. Even if this plan did work flawlessly, it would only work for a few hours until the sun rose to reveal the carrier. This tactic cannot be used against a competent enemy. Overall, visual scouting for surface warships is more effective today than ever before.

Radar

Radar changed naval and air warfare in World War II, but many radar systems were primitive and faulty in those formative days. Nonetheless, radar still permitted the early detection and targeting of the enemy. For example, in the Battle of Vella Gulf, Cdr. Frederick Moosbrugger used his radar to detect four Japanese destroyers and quickly launch torpedoes, sinking three ships.[32]

Radars have made incredible advancements in range, accuracy, and overall capability; once limited by the curvature of the earth and confined to large warships, radars are now built using technologies like phased-array and skywave systems to see farther and more precisely then ever possible before. These radars, mounted on everything from small missile craft to airborne warning and control system (AWACS) aircraft to UAVs, make it incredibly hard to hide a large target. Simultaneously, the target itself has only gotten bigger, offering a larger radar cross section. Jamming and decoy technologies can degrade some radars, although the ability to prevent the use of radar by dozens of platforms across all frequencies would require enormous resources and is highly questionable. Radar-absorbent materials and stealth hull designs help, but it is impossible to hide a 100,000-ton target from modern radar. Thus, just like the visual method of ship detection, radar is only making it easier to track warships at sea as time progresses.

Acoustic

The progression of acoustic technology has also been decidedly in favor of the archer over the last hundred years. Sonar was unavailable in the Age of Sail and, like radar, was primitive in World War II. Its limited range and sensitivity meant it was more often used to coordinate an attack than for initial detection and tracking. Sonar technologies have made dramatic advancements and now allow warships, submarines, and seabed hydrophone nets to detect targets at long ranges. Enemy submarines, using advanced spherical and cylindrical bow-mounted sonars coupled with towed arrays for long-range detection—unintentionally aided by U.S. escorts blaring away with active sonar pulses—could use sonar to detect and close with an aircraft carrier. Even if a precise solution cannot be developed, a submarine's report of the approximate bearing and location of a carrier strike group could be used as a starting point for other assets to search.

When reviewing the changing acoustic profiles of surface warships over time, the record is mixed. Ships are designed to be quieter and employ technologies like Prairie Masker, which uses a sheath of air bubbles along the hull and propeller to mask emitted noise.[33] However, warships are getting larger and tend to operate at high speeds, meaning more energy is being emitted into the water for enemy sonar to detect. In addition, the active sonars on U.S. escorts are much more powerful than in World War II; while increased strength increases the likelihood of detecting enemy submarines, it also greatly increases the range that those signals travel, thus providing another means of tracking the carrier strike group. All things considered, it is decidedly easier to detect and track a surface warship using sonar today than it ever has been in naval history, giving the archer another potential tool.

Human Intelligence

Unlike radar and sonar, human intelligence has been used since the Age of Sail to aid in the tracking of the enemy. During the War of 1812, American frigate captains often hailed passing friendly merchant ships to gain the latest intelligence. In the Pacific theater during World War II, coast watchers provided invaluable intelligence on Japanese air and naval movements. These coast watchers—including stranded servicemen, civilians, and locals—warned of Japanese activities using crude radio sets that could require up to sixteen people to transport.[34] Human intelligence is now an even more capable method of ship detection, with important changes both ashore and at sea. First, there are simply more people in the world today than during the Age of Sail or World War II, meaning there are significantly more eyes capable of sighting a fleet if it sails close to land. From approximately 1.1 billion people during the War of 1812 to 2.3 billion during World War II, the world population exceeded an estimated 7.7 billion in 2020.[35] Communications have enabled huge portions of these people to be potential intelligence sources. Instead of a strait having one coast watcher using a primitive radio, that same strait today has thousands of people all with cameras, cell phones, and social media access to instantly broadcast to the world that they just saw an American warship passing by.

Just as there are more people watching from the shore, there many more ships at sea as well. Two hundred years ago, a fleet attempting to stay hidden had to avoid the few other ships under way, and even if they made contact it could be weeks before that merchant was able to report its intelligence. Today, there are more than four million fishing boats and small merchants and more than 100,000 large merchant ships scattered around the globe.[36] Furthermore, instead of waiting weeks to pass on their intelligence, they can simply use cell phones or radios to communicate it, accompanied with precise GPS coordinates and pictures. Worse, some militaries can discreetly equip and train civilian ships to perform scouting functions, such as the Argentinians did in the Falklands Conflict or the PRC does today with its People's Armed Forces Maritime Militia (PAFMM) of trawlers and merchants.[37] Overall, in a densely populated and interconnected world, human intelligence is better able to find and communicate warships' locations today than ever before.

Electronic Warfare

Electronic warfare, for this study, can be loosely defined as the use of detected electromagnetic (EM) energy (including communications and radar) to locate an enemy. Unused in the Age of Sail, it gained prominence in World War II in a variety of forms, and was used effectively in numerous theaters to detect, track, and attack enemy assets. German submarines communicated extensively with Admiral Karl Dönitz, allowing the Allies to triangulate the U-boats' approximate location using high-frequency direction finders (HF/DF or "huff-duff") and attack them.[38] In addition, just as active sonar gives away the position of whatever platform is using it, warships were at times wary of using radar in the Pacific theater for fear of revealing their location to the very enemy they were searching for.

Today, there have been major changes in the field of electronic warfare and its use to locate the enemy. Benefiting the hunted, communications no longer have to be omnidirectional but can be directed at friendly assets or satellites while emitting little or no energy in directions that would allow an enemy to detect them. Even if radar or communications signals are detected, there are exponentially more signals being transmitted today, and determining which waveform belongs to a warship could be a challenge. Furthermore, ships have the option to employ emissions control, in which they minimize or eliminate energy transmissions off the ship. They can instead rely on supporting assets, such as the radar on a nearby AWACS plane.

However, many other changes benefit the archer. There is a great deal of evidence that operating with "radio silence" would be a tall task for today's carrier strike group. The billions of dollars of systems that are the strength of U.S. escorts, including the SPY-1 phased-array radar and the CEC, do not provide any value when turned off. When they are on, these systems and radars are immensely powerful and often operate at frequencies and with characteristics unlike those of any other system, making them quickly identifiable. Former Chief of Naval Operations Adm. Jonathan Greenert identified this problem, saying that the service must understand and improve its "electromagnetic hygiene."[39] Two lieutenant commanders—DeVere Crooks, a surface warfare officer; and Mateo Robertaccio, an intelligence officer—expanded on the CNO's comments:

> While we became fairly proficient at emissions control during the Cold War, the lack of a meaningful blue-water threat since the fall of the Soviet Union and our vast accumulation of new EM systems have allowed us to forget. Today, naval vessels and aircraft operate by default with multiple active radar, identification, datalink and communication systems radiating. Rarely are we forced to operate in a silent (or reduced) mode for any sort of extended period or while conducting complex operations. . . . In air defense, we so infrequently practice single-ship or group-restricted radiation operations in tactical scenarios that it is questionable whether we are proficient enough to use them in wartime. And, as an increasing number of antiship weapons no longer rely on easily detectable active-radar seekers, would we be confident enough in our tactics and equipment to leave our radars off while under threat of attack from weapons that our passive sensors might not see? Only practice and familiarity can breed that kind of confidence.[40]

Thus, with no obvious trend, with many of the technologies classified, and with so much dependent on the carrier strike group's operating posture, it is difficult to determine if searches using electromagnetic emissions are more or less effective today than during World War II.

Signals Intelligence

Signals intelligence, or the interception and decoding of enemy communications, has been used extensively throughout the history of warfare. During the

Age of Sail, captured mail often provided insight into the enemy's concerns and intentions. In World War II, the Allies cracked the German Enigma code to read their communications, leading to effective searches and attacks on U-boats. In the Pacific, the Americans also read large portions of the Japanese codes. Most famously, they were able to learn of the impending attack on Midway and the Japanese fleet's approximate location, using that information with smashing success.

It is unclear to what extent a U.S. carrier could be located using signals intelligence today. It is unlikely that foreign cryptologists could decipher the U.S. Navy's communications, yet the age of cyber warfare creates new vulnerabilities. There are numerous cases of hacking of U.S. government systems, including the alleged Chinese theft of more than 21 million records of individuals from the Office of Personnel Management in 2015 and the Russian penetration of multiple U.S. Federal departments and corporations in 2020.[41] Due to the secrecy, classification, and ambiguity inherent to signals intelligence, it is unclear whether the archer or the hunted is gaining the upper hand.

Satellites

The final major means of tracking a ship is satellites, and this means has obviously not favored the carrier. Today there are more than 2,700 operational satellites orbiting the planet, employing a variety of sensors including electro-optics, infrared, and radar.[42] Conversely, the carrier's ability to hide is doubtful at best. With a five-acre flight deck, accompanying escorts, a huge radar cross section, and a large infrared signature, there are few ships on the seas more recognizable than an aircraft carrier when surveilled from a satellite.[43]

Given all of these scouting methods—visual, radar, acoustic, human intelligence, electronic warfare, signals intelligence, and satellites—it appears that the overall historical trends in scouting methods do not favor those hoping to remain hidden. In both the Age of Sail and World War II, ships were consistently located, identified, and attacked despite their best efforts to hide. The men of the sailing age accomplished that tracking using only visual means along with slow human and signals intelligence. In World War II, the archer hunted using visual means, aided by primitive radar and sonar, human intelligence, and extensive electronic warfare and signals intelligence.

FIGURE 5.1. Historical Trends in Scouting Methods

Detection Method	Age of Sail	World War II	Today	Warship Susceptibility Changes	Trend
Visual	Primary means of detection	Primary means of detection, used extensively by scout planes	Used by planes and UAVs, used less by surface warships for initial contact	Larger, more recognizable ships	Harder for ships to hide
Radar	Unavailable	Primitive, but used extensively	Advanced radars carried on ships, planes, UAVs	Much larger radar cross sections, partially mitigated by jamming, stealth designs, and radar-absorbent materials	Harder for ships to hide
Acoustic	Unavailable	Primitive, not used for initial detecting and tracking	Advanced sonars and towed arrays carried on numerous platforms capable of detection at long ranges	Larger ships at higher speeds tend to be louder. Escorts' active radar use is another liability. Partially mitigated by quieting technology	Harder for ships to hide
Human Intelligence	Used extensively, but limited by slow speed of communications	Used extensively, limited to those with communication capabilities	More people living by the sea, often with access to cell phones and the internet to report ship sightings	Warships are becoming larger and more recognizable	Harder for ships to hide
Electronic Warfare	Unavailable	Used extensively. HF/DF used to locate numerous enemy ships	Better technology to detect and analyze signals	More energy (communications, radars, datalinks, etc.) emitted, although ship can operate using emissions control	Unknown
Signals Intelligence	Used intermittently, but limited by slow speed of communications	Used extensively. Allies broke both German and Japanese codes.	Difficult to crack coded communications, although there are cyber threats	Unclear due to classification issues	Unknown
Satellites	Unavailable	Unavailable	Many countries have satellites capable of at least partial carrier targeting.	Carriers are susceptible to satellite detection (recognizable, large radar cross section, large IR signature).	Harder for ships to hide

Modern navies can employ visual searches, advanced radar and sonar, human and signals intelligence, electronic warfare measures, and satellites to search for ships that have only gotten larger and more recognizable. These huge ships are more easily tracked today than at any point in naval history and the notion that a carrier strike group could stay hidden for long seems to be wishful thinking.

Many prominent officers and naval analysts have acknowledged the problems associated with relying on ships with massive signatures. Admiral Greenert wrote, "The large number, range of frequencies, and growing sophistication of sensors will increase the risk to ships and aircraft—even 'stealthy' ones—when operating close to an adversary's territory. Continuing to pursue ever-smaller signatures for manned platforms, however, will soon become unaffordable."[44] Cdr. Phillip Pournelle, a surface warfare officer and former operations analyst in the Department of Defense's Office of Net Assessment, has written several articles about the weaknesses of a fleet led by carriers with large signatures. In one he wrote, "Ships will be lost in the confusing and confined littoral region. The question is how much of a proportion of the U.S. Navy's capabilities will be lost when it occurs. Numbers matter, and having a large signature is a good way to get hit."[45] In another piece, Commander Pournelle continued his discussion of the aircraft carrier's massive signature. He wrote, "You can see them coming and you can tell where they are. There is an old story that during the run-up to conflict, the president routinely asks: 'Where are the carriers?' But now there is a new corollary: our opponents are asking the same question, and for them the answer is also the key to the knowing the whereabouts of the fleet."[46] Finally, esteemed analyst Norman Polmar, author of numerous works, including the multivolume *Aircraft Carriers: A History of Carrier Aviation and Its Influence on World Events* as well as *Ships and Aircraft of the U.S. Fleet*, declared, "Aircraft carriers have long been dependent on mobility for their survival, with their mobility making it difficult or impossible to pre-target them. Carriers are now vulnerable to continuous tracking by satellite and long-endurance unmanned aerial vehicles (UAVs). Related to this situation, modern carrier operations cannot be conducted in an electronically quiet or 'emission control' environment, which, coupled with the energy that a nuclear carrier puts into the water when she moves, makes her impossible to hide."[47]

The PRC has capitalized on this shift in scouting technologies. In space, the PRC has more than 120 military and civilian satellites that the U.S. Department of Defense reports "could support situational awareness of regional rivals and potential flashpoints, while monitoring, tracking, and targeting an adversary's forces."[48] In the sky, the PLAAF continues its rapid production of KJ-500s—its most advanced airborne early warning and control aircraft—along with hundreds of UAVs.[49] And at sea, its navy, the largest in the world, benefits from other key forces able to provide scouting data, such as the PAFMM. That fleet, likely consisting of hundreds of small vessels, trawlers, and merchant ships, provides a distributed network spread throughout the Chinese near seas that can camouflage with thousands of civilian vessels while discreetly providing targeting data on enemy forces.[50]

The PRC's most difficult scouting problem is likely the targeting needed to guide an ASBM onto a moving target at sea. However, they appear to have overcome even that challenge years ago; in 2011, Deputy Chief of Naval Operations for Information Dominance Vice Adm. David Dorsett said that China "likely has the space-based intelligence, surveillance, and reconnaissance (ISR), command and control structure, and ground processing capabilities necessary to support DF-21D employment . . . [and also] employs an array of non-space based sensors and surveillance assets capable of providing the targeting information."[51] The ability for carrier strike groups to operate close enough to China to have an impact while remaining hidden from the PLA's formidable reconnaissance network appears improbable.

This review of scouting trends indicates, in general terms, that it is easier to find surface ships at sea today than at any other point in naval history. The archer has more advanced tools with better global coverage, each of which can operate independently or cooperatively, with data flowing between sensors and commanders at incredible speeds. Conversely, the hunted has changed little since World War II, continuing to operate large, non-stealthy warships with distinct visual, electromagnetic, and acoustic signatures. The hope that the massive nuclear-powered aircraft carrier, radiating on numerous systems and accompanied by multiple large escorts, will be able to remain hidden for any useful period of time against a peer competitor seems implausible. In the carrier strike group's efforts to defend the aircraft carrier, "hiding from the archer" is not a viable option.

SHOOTING THE ARCHER

If the carrier and its escorts cannot remain hidden, the focus shifts to "shooting the archer" before he launches his arrow. This strategy refers generally to the attempt to attack or disrupt the enemy's command, control, communications, computers, intelligence, surveillance, and reconnaissance (C4ISR) systems or to destroy the ships, aircraft, and other platforms before they can launch their weapons. By examining the U.S. military's ability to attack the C4ISR infrastructure as well as ASCM and ASBM launch-capable platforms, it becomes clear that there are a lot of holes in the plan. It would be convenient if the U.S Navy was the only one shooting in a future war, but as the saying goes, "the enemy gets a vote." Depending on them to not be able to strike back is not a reliable strategy.

If the United States were to target the Chinese command and control organization, it would undeniably have many powerful weapons to employ in the effort. Tomahawk missiles fired from dozens of submarines and surface ships, guided munitions from F/A-18s, F-35s, and U.S. Air Force bombers, and cyberattacks would certainly cause great damage to a variety of targets and could greatly hamper enemy C4ISR operations. In the Persian Gulf War, American forces methodically destroyed much of the Iraqi communication, transportation, and defense systems before they could react, greatly reducing the threat facing coalition troops. Yet there are many reasons why this option may not be so attractive. Political leaders may be hesitant to authorize attacks on the Chinese mainland and could potentially restrict combat to the seas and cyberspace. Britain avoided attacks on mainland Argentina in the Falklands Conflict and the United States avoided direct attacks on China during the Korean War, despite engaging Chinese soldiers on the ground.[52]

If attacks are authorized it will be no simple matter to effectively destroy the necessary targets. The Defense Department assesses that within 300 nm of its coast, "China has credible Integrated Air and Missile Defense . . .

that relies on robust early warning, fighter aircraft, and a variety of SAM (surface-to-air missile) systems as well as point defense designed to counter adversary long-range airborne strike platforms."[53] This conclusion was supported by the RAND Corporation's analysis of the Chinese military, which stated, "The PLA has turned its air defense network from a flimsy distraction into a robust network that can successfully safeguard its airspace against all but the most advanced technology and tactics."[54] Much of the destruction of Iraq's infrastructure was possible because there was no credible threat to the coalition fleet or air force bases launching those attacks. A major power is not likely to let those carriers and bases go unmolested while they launch their attacks. Furthermore, sheer numbers will have an important effect. Compared to recent foes such as Kosovo and Iraq, China has a great many more targets scattered throughout a country that is so massive many U.S. weapons could not reach them.

In the quest to destroy the platforms launching ASCMs, the United States would again encounter serious problems. Whereas U.S. Navy firepower is centrally located in the aircraft carrier, PLAN ASCM firepower is spread out over numerous classes of ships, patrol craft, submarines, aircraft, and UAVs. Many Chinese ASCM variants outrange the primary U.S. weapon, the Harpoon, allowing the PLA to launch its weapons before being threatened by American cruisers and destroyers. Those PLAN warships have their own impressive antiair defenses that would help protect them against fighter attacks. Thanks to their small size, impressive speed, and sheer numbers, missile craft—led by the Houbei class—could prove very problematic to destroy before they launch their missiles. The plain fact is that the Chinese, as well as many other nations, have advanced surface warships that would be able to launch a credible attack on a carrier strike group. Even the Iranian attack boat *Joshan* was able to fire a Harpoon and nearly hit USS *Wainwright* during Operation Praying Mantis in 1988, despite attacking alone against an entire surface action group supported by an aircraft carrier.[55] Submarines are much more difficult to detect and attack before they launch their ASCMs and are addressed in Chapter 7.

As we have seen, the United States would have difficulty preventing one of the world's largest and most advanced air powers from launching ASCMs. Many of China's land-based fighter-bombers have longer ranges than the carrier-based American fighters, and the ASCMs that would be launched at the American fleet would likely outrange the defensive Aegis system missiles, meaning enemy aircraft could fire their weapons while minimizing their own risk. It is believed that American fighters are more advanced and operated by better-trained pilots than those of its peer adversaries, but the advantage is not so great that the United States can expect complete domination of the skies. In a war in China's near seas, this problem becomes even more difficult due to the limited quantity of fighters available. A typical carrier air wing has only forty-four fighters compared to the hundreds that are stationed along the Chinese coast.[56]

The combination of the growing capability of America's adversaries, the sheer number of platforms that are able to fire ASCMs, and the basic uncertainty of war make the idea of shooting the ASCM archer first a dubious proposition. Naval War College professors Yoshihara and Holmes wrote, "It is high time for naval officers to discard their shopworn assumption that high-tech warships or carrier aviation can strike down the 'archer' before he looses his 'arrow' at the fleet. This is a worthy goal, to be sure, but it is not a foregone conclusion."[57]

If the proposition of shooting most of the ASCM-shooting ships, submarines, and planes is unlikely, then the idea of destroying ASBMs and their launchers before they are fired is next to impossible. Chinese ASBMs are fired from a transporter-erector launcher (TEL), which closely resembles a typical tractor trailer when observed from the sky. Their size and mobility make them exceptionally difficult to find in a nation with an estimated 240 million cars and trucks providing camouflage.[58] The DF-21D's estimated 1,500-km range means that the Chinese could fire at a carrier strike group before many of the fleet's weapons are available. In addition, the short time it takes to set up the TEL for launch would make it exceedingly difficult for the United States to locate the missile, get word to a nearby submarine, ship, or aircraft, and complete an attack—all in time to destroy the ASBM before it is launched. Furthermore, the Chinese mainland is well defended and the political situation may prevent striking inland targets in any case.

In World War II the Allies controlled the skies but could not completely stop the launch of German V-1 and V-2 rockets. The dreaded rockets were often fired from mobile launchers, and Allied ground troops finally ended their reign of terror by overrunning the launch sites late in the war. In the Persian Gulf Wars, coalition forces were often unable to find Scud launchers, despite having air superiority.[59] In 2001 the RAND Corporation published a study of the U.S. ability to destroy elusive ground targets. It concluded that it was unlikely they could be found while in hiding and that it was "near impossible" to defeat Iraqi TELs if they operated in populated areas where they could hide in the heavy traffic.[60] In the same study, the authors used simple math to illustrate the immense difficulty the United States would have trying to destroy the TELs after they launched their missiles, never mind preventing the launch altogether. While the authors focused on theater ballistic missiles (TBMs) rather than ASBMs, and some of the exact figures are outdated due to the age of the report, the overarching theme is the same: effectively destroying TELs to "shoot the archer" is nearly impossible. The study extrapolates:

> Let us assume a 1-minute (min) delay from launch detection to compute and transmit the launch coordinates to an attack aircraft. Let us also assume that the TEL requires 5 min to tear down and move out of the immediate launch position or hide. This leaves 4 min for an attack aircraft to respond to a launch. An aircraft flying at Mach 0.8 (typical tactical aircraft cruise speed) can fly 32 nautical miles (nmi) in 4 min. Thus, each attack aircraft could, in theory, effectively patrol an area of about 11,000 sq km. Therefore, to patrol just the 320,000-sq-km TBM operating area near Taiwan would require about 30 attack aircraft on-station at all times. It would take about four U.S. attack aircraft based on Taiwan or Okinawa to maintain each orbit. Therefore, about 120 attack aircraft (almost 2 full fighter wings) would be required just to maintain the 30 anti-TEL orbits. These aircraft would need support from tankers, AWACS, and other aircraft. Attempting similar coverage of the cruise-missile TEL operating area would require a further 150 orbits—many of them almost 1,000 nmi deep inside China—and hundreds of additional attack aircraft to sustain a presence in these orbits. Of course, none of this would be possible unless the attack aircraft could safely orbit deep inside Chinese airspace, which would require the elimination of the advanced

SAM threat—a tough task that could take weeks to months to complete. Meanwhile, Chinese TBM TELs could operate with virtual impunity.[61]

Whether the target is an ASCM or ASBM, it is highly improbable the United States will be able to consistently shoot the archer before he launches the arrow. There have been isolated skirmishes and battles throughout history in which one side got very few or no shots off, but these are few and far between. The notion that a competent adversary would go an entire war without launching effective attacks is questionable at best.

SHOOTING THE ASCM ARROW

After the enemy has found the carrier strike group and fired ASCMs at it, the focus shifts to the fleet's ability to shoot down or divert the incoming weapons, or "shooting the arrow." The U.S. Navy's forbidding layered defense can be broken down into three stages: an outer ring formed by fighters, an inner ring formed by SAMs, and a final ring comprising point defenses, electronic countermeasures, and decoys. These actions are coordinated in part by the Aegis combat system as well as the E-2 Hawkeye aircraft.

The Aegis combat system is the world's premier antiair warfare system and is employed aboard *Arleigh Burke*–class destroyers and *Ticonderoga*-class cruisers, as well as the ships of several Allied nations. Aegis is designed to incorporate multiple systems, such as radar, fire control, and weapons launch all into one structure. A central feature of the system is the SPY-1 radar, which uses phased arrays to accomplish both search and fire control functions in a single system at speeds and with accuracies unheard of for many other platforms.[62] It operates using a wide variety of frequencies, is resistant to electronic jamming, can follow the ship's defensive weapons and provide mid-flight guidance, and it can track hundreds of targets out to ranges in the hundreds of miles.[63] Aegis also includes the Cooperative Engagement Capability, intended to allow nearby planes and ships to combine each unit's radar and fire-control data into one composite picture allowing for early defense against incoming threats.

Above the fleet, the E-2 Hawkeye airborne early warning aircraft serves to give the ships below a bird's-eye view of everything in the area. Equipped with a huge saucer-like radome, the different variants of the E-2 can track thousands of targets at incredible ranges and is considered the most capable AWACS aircraft in service today.[64] The data provided by the E-2 can be fed into the Aegis system or used to vector in fighters from the carrier. This gives the fleet unprecedented battlespace awareness.

The first ring of carrier defense is its air wing. In addition to its helicopters, E-2 Hawkeyes and single squadron of electronic attack fighters, a typical carrier will be led by four squadrons of strike fighters. These four squadrons have a total of approximately forty-four F/A-18s or F-35s.[65] The F/A-18 Hornet has been the workhorse of naval aviation for years and was designed to accomplish roles such as fleet air defense, close air support, and electronic warfare. By performing these functions, the F/A-18 replaced several specialized platforms such as the F-14 Tomcat and EA-6B Prowler. The F-35 Lightning Joint Strike Fighter, the Navy's first fifth-generation fighter, is now joining the fleet. Incorporating the latest in weapons technology and with a very small radar cross section, the F-35 embodies the future of carrier aviation. Both the F/A-18 and F-35 can carry a variety of weapons, but the AIM-120 advanced medium-range air-to-air missile (AMRAAM) is their primary means to shoot down ASCMs, although the missile was designed for use against enemy aircraft.[66] In aerial combat, very few if any fighters in the world outclass the F/A-18 and F-35. However, their performance against sea-skimming anti-ship cruise missiles is unknown, especially against modern supersonic models. No fighter has ever shot down an ASCM. Their most likely chance of success comes in shooting down the enemy attacker before the missile is released.

The next ring of defenses is composed of a variety of SAM systems carried on several different ship classes. The main defensive weapon is the Standard Missile (SM) family, including the SM-2, SM-3, and SM-6 interceptors, fired from the Mark 41 vertical launch system (VLS) aboard the carrier's escorts. The SM-2 is the primary missile for fleet defense; it is guided by inertial navigation, receives mid-flight updates from Aegis, and relies on semi-active radar or infrared for terminal homing.[67] The SM-6 was designed using parts of the SM-2 along with the seeker radar from the AMRAAM. It is intended for defense at extended ranges and is capable of speeds in excess of Mach 3.[68] The Standard Missiles are also intended for ballistic missile defense (as discussed in the next section). In the short- to mid-range, the Evolved Sea Sparrow

Missile (ESSM) uses semi-active homing, mid-course guidance updates, and supersonic speed to attack incoming ASCMs.[69] It is found on most surface escorts, as well as carriers themselves. When loaded into VLS, a four-pack canister can be loaded in a single missile cell.[70] The final prominent SAM system is the Rolling Airframe Missile (RAM). The RAM is a short-range, supersonic, fire-and-forget weapon that uses the Army's Stinger missile's infrared seeker with the propulsion, fuse, and warhead from the Sidewinder air-to-air missile.[71] It is limited to a range of approximately five nm.[72]

The final ring of ASCM defenses include electronic countermeasures, decoys, and point defenses. Electronic countermeasures, such as the jamming of an incoming missile, may be one of the most effective means to defeat an inbound threat. While current U.S. capabilities are difficult to assess due to classification issues and limited testing data, in the 1973 Arab-Israeli War both sides fired a total of approximately a hundred Styx and Gabriel missiles, yet the Israelis defeated all incoming Styx missiles using "soft-kill" measures and decoys.[73] Carriers and their escorts are currently equipped with the MK 53 Nulka Decoy Launching System, which radiates a large radar signal to simulate a ship for the ASCM to target and attack.[74] If all else fails, the Phalanx CIWS is the last line of defense. A self-contained automatic system that includes its own radar, the CIWS has a 20-mm Gatling gun capable of firing up to 4,500 rounds per minute, depending on the variant. This famous system tracks its own shots for aiming corrections, is designed to operate without human input when in automatic mode, and is on all U.S. carriers, cruisers, and destroyers (among other platforms).[75]

These rings of a carrier strike group make an American aircraft carrier the best defended surface ship in the world. There are numerous ships, planes, sensors, and weapons designed to protect the carrier, operated by the best trained and most dedicated officers and sailors the world has to offer. However, these defenses also have key flaws and have not been tested in combat in more than seventy years. Most importantly, when defending the centerpiece of American power, the defense cannot be merely "great"—it must be perfect *every time*, an achievement that is highly unlikely.

In the outer ring, the ranges of the defending fighters have been decreasing over time (1,600 nm to 1,200 nm for the F-14 and F-35, respectively) as the

Navy has shifted its focus from sea control and blue-water combat to power projection ashore against weak enemies.[76] Furthermore, despite the high quality of the fighters, they could very well lose the numbers game. Forty-four carrier fighters—even with their quality advantage—face daunting odds against hundreds of land-based fighters. Finally, China is developing stealth fighter-bombers, like the fifth-generation, long-range Chengdu J-20 which may prove difficult to detect, track, and destroy before it can launch its payload.

SAM systems have inherent disadvantages and unproven records as well. Most SM and ESSM series evaluations have been conducted against a single ASCM under ideal conditions using an alerted crew, limiting the tests' robustness. The director of operational test and evaluation (DOT&E) is the principal adviser to the Secretary of Defense for the testing and evaluation of military weapon systems. For several years the DOT&E has called on the Navy to build a Self Defense Test Ship (SDTS) in order to effectively evaluate the ability of Aegis, Standard Missiles, ESSM and other systems to defend a ship that it cannot test using a normal ship due to safety concerns. In 2014 the DOT&E announced that the operational test program for multiple defensive systems was "not adequate to fully assess their self-defense capabilities." When the Navy attempted to justify why it did not need to test its systems using an SDTS, the DOT&E found that the study included "a number of flawed rationales, contradicted itself, and failed to make a cogent argument for why an SDTS is not needed for operational testing."[77]

Just as there is little testing data, there are few combat successes to review. Only once in the fifty years of their existence has an ASCM been shot down by an interceptor missile, a rare success achieved by the HMS *Gloucester* against an old, slow Silkworm missile fired at USS *Missouri* during Operation Desert Shield.[78] This lack of combat data is part of the reason Captain Hughes wrote in *Fleet Tactics and Coastal Combat* that Aegis defenses' success and failure rates are "speculative and debatable."[79] Even if the interceptors do achieve incredible rates of hit success, the escorts can only carry a limited number of them and may quickly be forced to withdraw to resupply.

Finally, there are questions with the Navy's much-vaunted CIWS as well as other short-range defenses. In the 1987 Exocet attack on USS *Stark*, no radars detected the incoming missiles, the CIWS did not fire due to being in standby mode, and the SRBOC system did not actuate because it was not armed.[80] In a 2013 exercise the cruiser USS *Chancellorsville* was tracking an incoming target missile drone that malfunctioned and failed to turn away and instead hit the ship, causing fires and requiring six months of repairs. Despite the CIWS system identifying the target, the operator could not get permission to fire in time to prevent the subsonic drone hitting the ship.[81] These incidents highlight the extremely short reaction times crews face, and how the smallest error in the system or with its operators—even against an alert crew expecting ASCMs—can result in failure.

Electronic countermeasures and decoys may prove more effective but have their own issues. Jamming an incoming weapon requires knowledge of its guidance system. Filling the airwaves with radar signals does not affect a weapon with an infrared seeker. As Commander Pournelle wrote regarding missiles, "Employing an extremely wide range of the electromagnetic spectrum and sophisticated techniques within each band of the spectrum, it will become extremely challenging to effectively decoy such systems, particularly if the intended target has unique characteristics and emissions."[82] In addition, decoys do not destroy the missile, leaving it to acquire a new target—this is what led to the sinking of the SS *Atlantic Conveyor* container ship in the Falklands Conflict. HMS *Ambuscade* successfully defended itself from an Exocet ASCM using chaff, only to have the missile veer into the *Atlantic Conveyor*, which was carrying much-needed fighters, helicopters, and other supplies.[83]

The carrier's defensive rings are formidable, but far from insurmountable. With limited peacetime testing and almost no combat experience, no one truly knows how those systems will perform against salvo attacks from a peer adversary. No matter how well they perform, they are unlikely to provide the impenetrable defense needed for the U.S. Navy's flagship and the iconic symbol of American might.

When ASCMs do penetrate those screens to impact their targets, the damage will likely be severe. The Exocet missiles that hit HMS *Sheffield* and USS *Stark* did not even explode, yet *Sheffield* sank and *Stark* had extensive fires and suffered thirty-seven deaths.[84] One of the themes in *Fleet Tactics and Coastal Combat* is that modern warships have poor ability to sustain damage due to their minimal armor and an overreliance on interceptor defenses. Rear Adm.

Walter Carter Jr., a former carrier commanding officer and president of the Naval War College, declared, "No matter how heavily armored they are, ships cannot expect to retain their combat effectiveness after hits from specialized antiship munitions, given the precision and lethality" of long-range precision strike systems.[85] In reference to an aircraft carrier, former vice chairman of the Joint Chiefs of Staff Adm. James Winnefeld was even more explicit: "One hit on that thing is a mission kill."[86] The Navy's ability to "shoot the ASCM arrow" is doubtful, and in an era of high-powered precision munitions the consequences of failure are grave.

SHOOTING THE ASBM ARROW

If the U.S. Navy cannot prevent the launch of an anti-ship ballistic missile, its primary means of defending its ships is with Standard Missiles, aided by electronic countermeasures and ship mobility. Ballistic missile flights are typically divided into three phases of flight: boost, midcourse, and terminal.[87] The boost stage features the launch of the missile and its ascent into space; due to its large infrared signature, the missile is observable by satellite. During the longer midcourse stage, the missile coasts in space toward its target at supersonic or hypersonic speeds. Finally, the missile reenters the atmosphere to hit its target during the terminal phase.

Destroying ballistic missiles during the boost stage is the most difficult due to the short window of time. Intercept during the terminal phase is unfavorable due to the proximity to the target and lack of room for error. Thus, U.S. Navy defenses are designed to detect and track the missile during the boost stage using satellites, UAVs, and radars, and then destroy the missile either in the midcourse stage using an SM-3 interceptor, or during the terminal phase with an SM-2 or SM-6 interceptor. Successful detection, tracking, and destruction of a ballistic missile is extraordinarily difficult. Closing speeds can exceed four kilometers per second and interceptors often must be fired using a "blind launch" technique in which they are guided by another asset's sensors. The entire time from launch to impact for a threat like the DF-21D is approximately ten to fifteen minutes.[88]

Ballistic missile defense (BMD)–capable Aegis destroyers, firing modified Standard Missiles, are the Navy's principal ballistic missile defense system and the primary means of defending an aircraft carrier from an ASBM. As of 2020, forty-eight cruisers and destroyers have been upgraded to be able to conduct BMD.[89] Due to ballistic missiles' exoatmospheric flight and supersonic to hypersonic speeds, conventional defenses such as fighters, ESSM, and CIWS are essentially useless.

As with ASCM defenses, ballistic missile defenses are always disadvantaged because of physics, sheer numbers, and the element of surprise. Those disadvantages, along with the incredible cost of BMD systems, have led to poor track records for many earlier defense systems. President Reagan's Strategic Defense Initiative, nicknamed the "Star Wars Defense," was deemed unworkable after billions of dollars were poured into it. During the Persian Gulf War the U.S. Army artificially inflated its Patriot missile success rate against Scuds by characterizing any event in which a "Patriot and a Scud passed in the sky" as an intercept.[90] When President George W. Bush moved to build the ground-based midcourse defense (GMD) system, the Union for Concerned Scientists said the system had "no demonstrated defensive capability." They went on to call the administration's claims of the system's reliability and effectiveness to be "irresponsible exaggerations."[91]

Today, Aegis and similar systems have a better track record, but likely not robust or effective enough to defend aircraft carriers from large ASBM attacks. As of 2020, SM-2, SM-3, and SM-6 Aegis defensive missiles had achieved successful intercepts in forty-one of fifty attempts against ballistic missile targets.[92] However, few if any tests involved more than one target, they often were conducted under ideal circumstances against an alerted crew of a specially trained ship, and decoys do not appear to have been used in any tests. Additionally, these tests have been against conventional land-attack missiles and not true ASBM targets. In 2012 the DOT&E said that there was no target that "adequately represents" an ASBM's trajectory, and that the Navy had not "budgeted for any study, development, acquisition or production" of such a target.[93] The Navy's progress, if any, since is unclear because subsequent DOT&E reports on the subject have been classified.

Some of the Aegis BMD system's most punishing and public criticism came in 2010 from George Lewis, a Cornell University experimental physics professor, and Theodore Postol, a Massachusetts Institute of Technology professor of science, technology and national security policy and former scientific adviser to the Chief of Naval Operations. The professors slammed the technical oversight and testing of BMD systems as "deeply flawed and unreliable," and declared that the rosy vision put forth by the Missile Defense Agency (MDA) was a "fiction."[94] Reviewing the MDA's test data, Lewis and Postol cited multiple testing parameters that were highly favorable for the defenders; for example, they noted that there was nothing else in the SM-3's search volume to distract it (no decoys). They added that the warhead had large fins and was always attacked with a side-on geometry, thus maximizing the radar cross section, and finally, they said that geometry of the target was always known. They concluded that the MDA's test data showed the missile defense systems to be "fragile and unworkable" and a "transparent bluff."[95]

Even if Aegis interceptors do prove highly accurate, the simplest means of overcoming the carrier's defenses is to just overwhelm them with numbers. As of 2015, the Navy had a total of 165 SM-3 missiles.[96] These missiles, along with the BMD-capable escorts, are spread across the globe. Even if half of them were at sea and operational in the western Pacific when war broke out, a very generous assumption (given maintenance and deployment schedules) is that it would only leave about 83 missiles to shoot down China's current estimated inventory of dozens to hundreds of ASBMs.[97] This defense is further complicated because those interceptors are needed for numerous other threats, as well; as Adm. Harry Harris testified when he commanded U.S. Pacific Command, China "controls the largest and most diverse missile force in the world, with an inventory of more than 2,000 ballistic and cruise missiles."[98] Even if the SM-2 and SM-3 missiles are able to achieve a hit every time, the battle group will quickly run out of missiles and be forced to withdraw. HMS *Sheffield*, HMS *Glamorgan*, and USS *Stark* all had modern air defense systems for their time, yet all three failed to defend against an attack of only one or two anti-ship cruise missiles. These ships should serve as warnings against the notion that today's missile defenses will be perfectly effective against dozens or hundreds of incoming missiles.

Further complicating the problem of quantity, the ASBM also has the advantage of flexibility. Making trajectory changes, adding decoys, and building more ASBMs is a lot cheaper than developing new interceptors and building more destroyers and cruisers to launch them, which is not a disadvantage that the United States can realistically overcome.[99] The ASBM attacker can react to the tactical picture and weather to launch an attack whenever desired, whereas the defender must launch interceptors with only minutes notice, and thus must always maintain a high state of readiness.

Finally, common sense and intuition must serve as guides when determining the relative likelihood of an ASBM being able to hit a carrier, and of an Aegis interceptor being able to hit the ASBM. In the simplest terms, who has the advantage: is it the attacker shooting the bullet, or the defender who must shoot that bullet with another bullet? Many carrier proponents argue that the Chinese will find it technically impossible to hit a carrier with a ballistic missile. Many of these same individuals fervently believe that the technology has been perfected in the Aegis defensive system to hit ballistic missiles with incredible consistency. However, looking at the two systems and their inherent differences challenges the validity of these conclusions.

The ASBM must hit a target that is moving at roughly thirty knots, has a surface area equal to five acres, and can only move in two dimensions. Aegis missiles must hit a target that is moving at roughly five thousand knots, is approximately one square meter in cross-sectional area, and is moving in three dimensions.[100] It is very difficult to use chaff to replicate the massive radar signature of a carrier, but there are numerous possible decoys that could be deployed to effectively replicate the radar and infrared signature of a DF-21D warhead. The Chinese can launch an ASBM on their own schedule, whereas the SM-3 interceptor must be fired within a few minutes of the initial detection of an ASBM launch and with little to no warning. Simultaneously arguing that the ASBM is technically impossible and that Aegis BMD is nearing technical perfection is contradictory; it is evidence of our natural human tendency to inflate our own abilities while diminishing those of our adversaries.

Capt. Jerry Hendrix, a retired aviator and a historian, drove home the difficulty of defeating the DF-21D, stating, "If there is a way to defend against it, I'm not aware of it."[101]

A BETTER WAY TO FIGHT THE ARCHER

Cdr. John Patch, a retired surface warfare officer, wrote that proponents of the aircraft carrier "universally seem to accept on faith alone the premise

FIGURE 5.2. Comparing the ASBM and the Aircraft Carrier as Targets

	ASBM Targeting Carrier	SM-3 Targeting ASBM
Target Speed	30+ knots	5,000+ knots (Based on Pershing II speed of Mach 8)
Target Radar Cross Section	20,000 m (5 Acres)	1 m (Based on Pershing II warhead dimensions)
Target Motion	2-Dimensional	3-Dimensional
Target Decoys	Chaff	Chaff, Mylar balloons, radar reflectors, infrared emitters, debris, additional warheads
Required Launch Time	None Can launch ASBM when ready	Must fire SM-3 within minutes of ASBM launch, likely with no warning

Table Notes
1. "Aircraft Carriers - CVN: U.S. Navy Fact File," U.S. Navy, last modified October 16, 2014, accessed August 12, 2015, http://www.navy.mil/navydata/fact_display.asp?cid=4200&tid=200&ct=4.
2. "U.S. Navy Sees Chinese HGV as Part of Wider Threat," *Aviation Week and Space Technology*, Jaunary 27, 2014, http://aviationweek.com/awin/us-navy-sees-chinese-hgv-part-wider-threat.
3. "Building a Giant: *Gerald R. Ford* (CVN 78)," Huntington Ingalls Industries, last modified October 11, 2013, accessed August 7, 2015, http://thefordclass.com/doc/Ford-fact-sheet.pdf.
4. "Pershing II Weapon System Operator's Manual" (Washington, DC: Headquarters, Department of the Army, June 1, 1986), http://pershingmissile.org/pershingdocuments/manuals/tm%209-1425-386-10-1.pdf.

that a nuclear-powered aircraft carrier (CVN) is essentially invulnerable."[102] This chapter aimed to test that faith by objectively examining the carrier's defenses and thus its ability to execute its missions. A review of the three main methods of defense—hiding from the archer, shooting the archer, and shooting the arrow—indicates that aircraft carriers are well defended but nowhere close to being invulnerable. The idea of continuously hiding five to seven large warships has been dispensed with, as has the notion of being able to prevent attacks from a competent adversary. Finally, the ability to consistently intercept supersonic maneuvering warheads with complex guidance systems, hurtling at the carrier at wave-top height or from space has been shown to be untested and unlikely. Taking these offensive approaches in conjunction, we can speculate that fleet defenses will certainly be able to stop a significant portion of attacks, but are unlikely to come anywhere near perfection. Assume that the carrier strike group can consistently achieve an impressive 95 percent probability of defeating an incoming missile. In that case, then if just a single

Chinese Houbei–class patrol craft fired its eight missiles, there would still be a 34 percent chance of a missile hitting the carrier.

The question then becomes, what if some attacks are successful? Major ships have been sunk or damaged and fleets have gone on to victory despite the loss, such as when Oliver Hazard Perry lost his flagship USS *Lawrence* in the Battle of Lake Erie yet went on to win the battle.[103] The difference is that today most of the fleet's firepower is located on one class of ship. The accompanying cruisers and destroyers are there primarily to protect the carrier, sacrificing much of their potential offensive firepower for defensive interceptors. If the carrier is sunk, much of the fleet's ability to function is lost, a problem only confounded by the extreme political implications of losing a ship with thousands of sailors. The loss of *Lawrence* in the War of 1812 was painful but could be overcome. The loss of a $12 billion vessel with five thousand people onboard is unfathomable. Captain Hendrix discussed the severe political implications of a damaged or destroyed carrier, arguing that administration officials would recommend against the president sending carriers into battle if there was even a 10 percent chance that they would be disabled or sunk:

> The loss of an aircraft carrier, with images of a thousand American dead, or just having it disabled, with all its airplanes and radars knocked out and huge gaping holes in it, is such a heavy political blow that we probably wouldn't risk it unless it was for the actual defense of the continental United States. So we've created an asset that we cannot afford to lose because it's become such an iconic symbol of American power that to have that symbol damaged or destroyed would undermine the legitimacy of America's role in the world.[104]

To attempt to prevent this unthinkable act from occurring, the United States has constructed an elaborate and expensive defensive system that attempts to defy physics and thousands of years of naval history. That defense—protecting a ship that cannot be lost—is assigned the impossible task of being perfect.

Adm. Stansfield Turner identified the inherent problem with defending such a ship, arguing that "technologies that make our forces more lethal will be available in time to others. When opponents acquire remote sensing and precision, long-range targeting capabilities, as they are bound to do, the huge detection signature of the hundred thousand tons of steel in one of today's aircraft carriers will be a tremendous liability."[105] Israeli rear admiral Yedidia Ya'ari also advised against relying on supposedly perfect defenses when he wrote in a 1995 *Naval War College Review* article, "The surface ships now in commission were designed with the open ocean and distant defensive perimeters in mind; to keep deploying them to a playing field where, under the most optimistic assumptions, their survival requires as a normal operating mode the highest level of *everything, all the time,* is unhealthy and unrealistic in the long run."[106]

The best solution to this problem is a force structure like the Flex Fleet. The Flex Fleet acknowledges the extreme technical difficulty and massive expense associated with defensive missile systems—shooting the arrow—and so shifts much of its defense to hiding from and shooting the archer.

The Flex Fleet's additions all have significantly smaller radar, visual, acoustic, and electromagnetic signatures than those of aircraft carriers, and so are each more difficult to track. Furthermore, a fleet of numerous smaller ships is overall much more difficult to track than a fleet concentrated around a small number of capital ships. The Flex Fleet could not stay perfectly hidden from a capable enemy, but would be harder to target than a carrier-centric force.

The Flex Fleet would also have an enhanced ability to shoot the archer. With more ships, aircraft, UAVs, submarines, and missiles, there would be increased ability to find, track, and destroy the enemy before they could conduct their own attacks. When the enemy conducts attacks, which the Flex Fleet concept acknowledges as a reality of war, those attacks will likely be against smaller, less vital corvettes and frigates, rather than the Navy's capital ships. Finally, if and when the enemy is able to destroy U.S. ships, the Flex Fleet would be better able to absorb those losses and still complete the mission. The loss of corvettes, frigates, and various other ships would be painful, but not crippling for the Flex Fleet. As Adm. Arthur Cebrowski and Stuart Johnson wrote, "*fleet* survivability not *individual ship* survivability is what dominates."[107]

The Flex Fleet is a not a perfect solution, just a better one. Its smaller, more numerous ships are not invisible, but they are moderately harder to find. Its diverse missile shooters will not always be able to shoot the archer first, but their offensive focus means they will be able to attack effectively first more often. The Flex Fleet will not be able to perfectly defend all of its ships, but when losses inevitably occur, they will have a smaller impact on the fleet's overall firepower and combat strength. Most importantly, the Flex Fleet is a better option than the carrier-centric fleet because it is held to a different standard. Today's fleet defenses are very good, but they must be perfect to defend the priceless aircraft carrier. The Flex Fleet's defenses are also very good, which is good enough for a fleet designed to survive losses and fight on.

If the carrier's defenses were perfect, then that fleet structure would be the best option for the Navy of today and of the future. However, a review of the threats present at sea today indicates a near-perfect defense is highly unlikely. In his 1997 award-winning U.S. Naval Institute *Proceedings* article, then-lieutenant David Adams called for the Navy to realize it was anything but invincible. He cited several examples of American hubris, including the words of the U.S. Atlantic Fleet commander in chief at the time. The admiral boasted, "I would hate to fight an American right now. You would lose so bad your head would spin." The admiral also argued that the United States was so powerful it could throw a shutout and win a war with no casualties: "If we have the technological advantages . . . ; if we have control of the information . . . and we control the time line, why don't we just pitch a shutout? You see the American people have put a standard on us that is good for us. They get zero [casualties]; you get the victory. That's good military thinking."[108]

If a baseball team wins one night because the pitcher and defense complete the rare shutout, it is a fortunate occasion the manager should relish. However, that manager would be imprudent to base his entire season's strategy on getting a shutout out of his pitcher and defense every night. Similarly, if the U.S. Navy can win future naval battles using a perfect defense that prevents the enemy from ever finding, attacking, or hitting the fleet, then that is an exceptionally fortuitous and desirable outcome. However, the Navy's leaders would be wise not to base their entire strategy for a future war on consistently accomplishing that unlikely feat.

CHAPTER 6

LESSONS FROM UNDERSEA WARFARE HISTORY

Battleships are the ships of yesterday, aircraft carriers are the ships of today, but submarines are going to be the ships of tomorrow.[1]

— Fleet Adm. Chester W. Nimitz, USN

On May 1, 1982, the British submarine HMS *Conqueror* reported that it had located the Argentinian cruiser ARA *General Belgrano*. While waiting for permission to attack, the nuclear-powered *Conqueror* stalked the Argentinian cruiser, along with her two Exocet-armed destroyer escorts, periodically coming to periscope depth to check on them and then going deep to use higher speeds to catch up.[2] After a day of trailing, *Conqueror* received updated rules of engagement that freed it to attack; she closed to the point-blank range of 1,400 yards and fired three unguided Mk 8 torpedoes. *Belgrano* and her escorts continued steaming, unaware of the British submarine less than a mile away and the torpedoes running toward her at approximately forty knots. Within a minute of launch, two torpedoes slammed into *Belgrano*, ripping open her hull and killing 368 Argentinians. As the cruiser sank, *Conqueror* sped off to open range, safely returning to periscope depth to observe *Belgrano*'s destroyers haphazardly dropping depth charges on a target they never detected that was already eleven miles away.[3]

Conqueror's attack was a key moment in the 1982 Falklands Conflict and in modern submarine history. Tactically, the attack removed one of the Argentinians' few capital ships. Strategically, it had a much larger impact; out of fear of additional submarine attacks, the entire Argentinian navy withdrew to port for the remainder of the war in a stunning acknowledgment of their lack of antisubmarine defenses.[4] Most importantly, the attack demonstrated the disproportionate effect submarines can have. In the only case of a nuclear-powered submarine engaging in combat, a single submarine launched a single salvo that sank a single ship and, in doing so, defeated an entire navy.[5]

While the previous two chapters examined air and surface warfare, the subsequent two chapters review undersea warfare to understand the aircraft carrier's ability to overcome submarines and mines to accomplish its missions. With most of the U.S. Navy's firepower centered on the aircraft carrier, can an enemy submarine sink a carrier, damage it enough to force its withdrawal, or even just deter it from entering the theater at all? In 1982, the Royal Navy sank a capital ship and in doing so defeated the entire Argentinian surface fleet. It is not unreasonable to query: if a Chinese or Russian submarine could defeat an aircraft carrier, would it stymie the entire U.S. surface fleet as well?

To answer these questions, this chapter relies on the lessons of undersea warfare; as naval strategist Alfred Thayer Mahan wrote, "The study of military history lies at the foundation of all sound military conclusions and practice."[6] Chapter 4 conducted a similar review of air and surface warfare, showing that what little recent combat has occurred indicates that the carrier's utility is waning. However, there is even less modern submarine combat; since World War II, there have only been three brief submarine engagements. This makes it exceptionally difficult to apply the lessons of history to understand the submarines' potential against capital ships.

As a result, this chapter uses a modified approach that starts with an analysis of submarine combat in World War II, using that large data set to establish a known starting point. The chapter then reviews the technological and tactical changes that have occurred since then, for both submarines and antisubmarine forces, to understand which side has advanced further since World War II. Using that starting point and the trends since, it is possible to infer which force has the advantage today. The chapter tests that hypothesis with a study of the few cases of modern submarine warfare. The chapter also studies modern mine warfare and its potential impact on capital ships.

Those reviews portend a challenging future for a ship that must succeed for the U.S. Navy to win in battle. In World War II, submarines suffered heavy losses to antisubmarine forces, yet were still able to inflict serious losses on warships. After the war, numerous technological and tactical changes in undersea warfare suggest that submarines have surpassed their ASW opponents in lethality and are capable of wreaking havoc in modern warfare. Similarly, mines have proven to be cheap, effective means of destroying, damaging, or deterring enemy capital ships and fleets; the post–World War II U.S. Navy has suffered more than four times as many ship casualties due to naval mines than to all other weapon systems combined.[7] Submarines and mines are a threat to all U.S. Navy ships, not just its aircraft carriers, but the key difference is that the U.S. Navy's success in battle is not wholly dependent on those other ship types. Designing a fleet reliant on a single ship type—the nuclear-powered aircraft carrier—is only an effective option if it can execute its missions without interference. History indicates that submarines and mines have good chances of interfering with the carrier's operations—thus giving the enemy a path to defeating the U.S. Navy.

SUBMARINE WARFARE IN WORLD WAR II

World War II is an excellent starting point for the study of submarine warfare because of its large and instructive combat dataset. Those undersea battles showed that, in general terms, capable ASW forces tended to defeat equally capable submarines, resulting in heavy submarine losses. For example, the Allies eventually won the hard-fought Battle of the Atlantic, destroying 642 of 859 U-boats that made at least one war patrol and killing approximately 29,000 of the 35,000 German submariners who saw combat.[8] In the Pacific, the Japanese devoted much less energy to antisubmarine operations, yet still proved effective enough to make the submarine force the deadliest major service of the war for American servicemen.[9]

To understand the submarines' potential against aircraft carriers today, we must grapple with the factors that led to their defeat in World War II. Why did submarines generally lose to equally matched ASW forces? What prevented them from sinking larger numbers of capital ships, and do those these negative factors still affect submarine warfare?

To answer these questions, we should emphasize that the submarine's most important attribute is *stealth*. Whether it was CSS *Hunley* in the Civil War, U-boats in World War II, or fast-attack submarines today, stealth gives the submarine the element of surprise and the option to engage in battle only if it chooses. Furthermore, the submarine's defenses are largely based on its concealment. When the enemy can reveal the presence and location of a submarine, its destruction is significantly easier. In his 1966 award-winning U.S. Naval Institute *Proceedings* article, "The Submarine's Long Shadow," Cdr. Robert Smith described this vital attribute: "What is it, then, that defines the submarine? It is not speed. There are ships that go faster. Nor is it weapons. The submarine possesses none that cannot be carried in other hulls. Least of all is its defensive strength. The submarine is a heavyweight, but it has a glass jaw. The unique attribute of the submarine, from which all its other virtues flow, is simply its ability to hide in the sea."[10]

Antisubmarine forces were able to inflict heavy losses on their submerged foes because of several important methods they could use to locate submarines, thus overcoming the stealth that was vital to submarines' survival and success. These vulnerabilities were a product of two aspects common to World War II submarines: their surfaced operations and their communications.

World War II submarines operated on the surface most of the time. When surfaced, they could run their diesel engines, charge their batteries, and operate at much faster speeds than when submerged. They also had drastically improved scouting ability because they could use radar and take advantage of significantly improved visual searches. When submerged, submarines lost a great deal of their mobility, endurance, and battlespace awareness, and so tended to do it only to escape enemy counterattacks or to conduct daylight attacks on protected targets. As a result, these ships were generally considered "submersible warships" and not true submarines.

However, when operating on the surface, submarines sacrificed their stealth and were vulnerable to two of their greatest enemies of the war—radar and aircraft. Radar and aircraft revealed the position of hundreds of submarines throughout World War II, leading to aborted attacks, rerouted convoys, and oftentimes the submarine's destruction. For example, of the 566 frontline U-boats sunk by the Allies while at sea, 51 percent were destroyed by aircraft acting independently or in conjunction with a surface escort.[11] In addition, of the 36 U.S. submarines known to be sunk by Japanese forces, more than 30 percent were due at least in part to aircraft attacks.[12] Aircraft, often operating with radars, could cover huge areas, find a surfaced submarine, and attack it before it could submerge. Grand Admiral Karl Dönitz, who directed German U-boat operations throughout the war, described the impact of these two forces in his war diary in 1943, writing, "The enemy radar location device . . . is together with enemy aircraft at present the worst enemy of the U-boat. . . . The location device is robbing the U-boat of its most important characteristic, its undetectability."[13] As the Battle of the Atlantic progressed, the Allies' improved radar, as well as the aircraft operating from various land bases and eventually escort carriers, forced more and more U-boat captains to either operate submerged and drastically reduce their offensive capabilities, or accept large amounts of risk to operate surfaced.

Communications, including the decryption of enemy signals and radio direction finding, provided another means for ASW forces to locate and attack submarines. Allied code-breaking efforts allowed the Allies to dispatch ASW forces to attack submarines that had revealed their location, and enabled commanders to route convoys around enemy submarines. After the war, U-boat historian Dr. Jürgen Rohwer concluded that as many as three hundred merchant ships were saved in the second half of 1941 alone by evasive convoy routing, enabled by the breaking of the German Enigma code.[14] Codebreakers in the Pacific enjoyed similar success, leading to the destruction of hundreds of Japanese submarines. For example, when the Japanese dispatched nine submarines to contest the U.S. invasion of the Gilbert Islands in late 1943, U.S. codebreakers tracked them all and enabled ASW forces to sink six of them.[15]

Radio direction finding also grew to be an important technology that revealed submarines' positions and negated their stealth advantage. High-Frequency Direction-Finding (HF/DF), which became known as "Huff-Duff," used receivers to detect the bearing to a transmitting submarine. When more than one HF/DF receiver was available, it was possible to triangulate the position of the submarine. Although these receivers were originally primitive and normally only land-based, by late 1942 most convoys had two or three HF/DF-equipped escorts.[16] When U-boats communicated with Admiral Dönitz and with other U-boats to coordinate attacks, escorts were alerted to the presence and location of the submarine and could strike before the U-boat closed the convoy.[17]

Radar, aircraft, codebreaking, and HF/DF all combined to rob World War II submarines of the stealth they so desperately needed. Yet even when they were able to remain hidden long enough to conduct attacks, faulty torpedoes resulted in numerous missed opportunities for both the American and German submarine fleets early in the war. American torpedoes were plagued with problems, including issues with their running depth and faulty exploders which caused numerous shots to result in duds or premature explosions. American submarines did not have an effective torpedo until September 1943, after almost two years of fighting.[18] In the Atlantic, German U-boats also had significant torpedo problems in the early stages of the war that resulted in numerous lost opportunities. At the end of the Norwegian campaign in 1940, Admiral Dönitz declared, "Torpedo failures cheated the boats of sure success," a statement backed up by British submariner Vice Admiral Sir Arthur Hezlet, who examined German logs and said that torpedo failures deprived the U-boats of "almost certain success against the [battleship HMS] *Warspite*, seven cruisers, seven destroyers and five transports."[19]

Thus, World War II submarines had numerous challenges facing them in any attempt to sink a capital ship. They were stealthy when submerged but needed to surface to conduct searches and to intercept enemy warships. When they did surface, they could be discovered by enemy aircraft or radar, or located using codebreaking or HF/DF. In the beginning years of the war, before many of these ASW technologies were reliable and widespread, submarines were often thwarted by faulty torpedoes that failed to run at the correct depth, did not explode after achieving a hit, or sometimes completed erratic runs to threaten the firing submarine.

Despite these serious difficulties, submarines achieved numerous scores on enemy warships to complement their massive successes against merchant shipping. For example, on November 13, 1941, U-81 penetrated the British escort screen submerged and hit the carrier HMS *Ark Royal* with a single torpedo, causing it to sink the next morning.[20] In the Pacific in September 1942, the Japanese submarine I-19 fired "undoubtedly the most effective torpedo salvo in submarine history."[21] I-19 penetrated the American escort screen, closed to within five hundred yards of the carrier USS *Wasp*, and fired six torpedoes. Three torpedoes hit *Wasp* which caused massive fires and its eventual sinking, while the other three missed. However, due to the incredible range of the Japanese weapons, those remaining torpedoes reached another task group five miles away. One torpedo hit the battleship USS *North Carolina*, causing significant damage, while another hit the destroyer USS *O'Brien*, causing it to sink while on its way for repairs.[22] During the Battle of the Philippine Sea in June 1944, the submarines USS *Albacore* and USS *Cavalla* sank the Japanese carriers *Taiho* and *Shokaku* within a few hours of each other.[23] Submarines scored unconventional successes as well, such as when U-47 sailed into the British base at Scapa Flow and sank the battleship HMS *Royal Oak*, or when Italian "human torpedoes" sank two British battleships and a tanker in the main fleet base at Alexandria in Egypt.[24]

These are just a few examples of submarines' many successes against warships, even when protected by competent escorts. Counting all sizes of aircraft carriers throughout World War II, aircraft of all nations sank twenty carriers of 342,000 total tons, and submarines were close behind, sinking fifteen carriers of 306,000 total tons.[25] Nearly one third of all Japanese combatants sunk were destroyed by U.S. submarines.[26] Thus, despite their many flaws and disadvantages, submarines inflicted heavy losses against capital ships and their escorts throughout World War II.

Despite these successes, it was clear submarines still had not reached their full potential, something that would only change with new technologies and tactics. Grand Admiral Dönitz promised this brighter future in a message to his U-boat captains, saying, "Shortly the day will come when, with new and sharper weapons, you will be superior to your opponent and will be able to triumph over your worst enemies, the aircraft and the destroyer. . . . Then we shall be victorious, my belief in our arm and in you tells me so."[27] That day did not come soon enough for the submarine of World War II, but many of the changes in submarine and antisubmarine warfare over the last seventy years indicate that Dönitz's prediction has come true. If the submarines of

World War II, which were limited in speed and endurance, typically operated on the surface, and had significant communications vulnerabilities, were still able to sink dozens of major warships, then it should be expected that modern submarines, lacking many of those vulnerabilities, likely pose an even greater threat to capital ships today. History does not appear to support the notion that ASW forces will be able to provide the perfect defense the aircraft carrier needs to execute its missions.

SUBMARINE WARFARE CHANGES SINCE WORLD WAR II

In the seventy years since World War II, both submarines and their ASW foes have benefited from numerous technological and tactical improvements. Submarines have profited from major advancements in propulsion and hull design, communications, and weapons, whereas antisubmarine forces have profited from new and improved detection methods, new weapons of their own, and the use of friendly submarines for ASW work. However, in general terms, the changes over the last seventy years have tended to benefit the submarine. Many of its greatest vulnerabilities, stemming from its surfaced operations and poor communications, simply no longer exist. The "submersible warship" of the 1940s has become a true submarine today, and there are very few ways for its enemies to defeat its stealth to attack it.

Submarines' advantage is far from absolute, but it is strong enough to make them another massive threat to the aircraft carrier. This evolution in undersea warfare points to a future in which it is extremely difficult and expensive to effectively protect the carrier when in harm's way. Just as with air threats, U.S. Navy leaders will be forced between risking a $12 billion ship with five thousand sailors or withdrawing the fleet to safe waters where the air wing is out of range. The aircraft carrier can be an impressive offensive weapon, but not if the ship is at the bottom of the sea or thousands of miles away from the fight.

The first major change benefiting submarines is in propulsion technology and hull design. Conventional submarines have an array of new technologies that allow them to stay submerged for extended periods while sailing deeper, faster, and quieter than ever before. Snorkels, extendable masts that allow submarines to draw in air for their diesels while remaining submerged, were used on German U-boats late in the war and are now standard in fleets worldwide.

When these submarines operate on their battery, they are extremely quiet due to the small amount of operating machinery. Furthermore, these batteries have grown in capacity since World War II, extending the time the ship can operate without running its diesels. When they are forced to recharge their batteries, sound silencing technology has substantially reduced the noise emitted, minimizing the ship's acoustic vulnerability. There are other technologies, like air-independent propulsion (AIP) found on ships such as the Chinese Yuan class, which significantly decrease the need to surface or snorkel altogether.[28] These advances have resulted in some of the stealthiest submarine classes in the world, such as the Japanese Soryu, Chinese Song, Australian *Collins*, and the Kilo, operated by numerous countries, including Russia, China, and Iran. Finding a surfaced U-boat using aircraft-mounted radar was challenging but possible in World War II. Finding a submerged Soryu, Song, *Collins*, or Kilo today is exponentially more difficult, even with advances in ASW methods.

The most important submarine development came on January 17, 1955, when USS *Nautilus* signaled, "Underway on nuclear power."[29] Nuclear propulsion fundamentally changed how submarines operated. No longer required to surface or snorkel to supply air for their diesels, and no longer limited in speed or endurance by their batteries or fuel, they could sail at speeds unheard of before. For example, the average fleet-type American submarine in World War II was capable of twenty knots on the surface and nine knots submerged for short periods of time, whereas today's U.S. submarines are capable of speeds in excess of twenty-five knots indefinitely.[30] Furthermore, these World War II submarines were capable of cruising ranges of approximately ten thousand miles, whereas nuclear-powered submarines are only limited by how much food they can carry. Historian John Keegan wrote, "The launching of USS *Nautilus* achieved a modernisation of the submarine revolutionary enough to imply that the aircraft carrier might not, after all, persist in its apparently ordained role as mistress of the twentieth century oceans. . . . *Nautilus*, in short, realised the dream of the submarine pioneers, being a true submarine and not merely a submersible boat."[31]

Another key submarine development was the teardrop hull design, first seen on USS *Albacore*. This hull form was designed for submerged operations, whereas all other previous designs had been focused on surfaced steaming. The teardrop hull has a smooth cylindrical form that significantly improves hydrodynamic effects to maximize submerged speed and minimize hull flow noise. This hull design and other engineering advances mean submarines can dive to deeper depths; for example, most U.S. submarines were limited to three hundred feet during World War II, but today's American submarine classes can dive to depths greater than eight hundred feet.[32]

These propulsion and design developments greatly improve submarine warfighting potential. By never surfacing, submarines are much less susceptible to surface escort and aircraft visual and radar searches that were so effective in World War II. Furthermore, submarines are no longer restricted to mediocre submerged speeds for short periods of time before slowing down to a crawl. Many successful submarine attacks in World War II occurred because the surface ships unknowingly sailed within striking range of the slow submarine, which capitalized on its luck. However, today's submarines are much less dependent on luck as they can use their speed to reposition and set up an attack, making them an even more capable foe.

Commander Smith, in the same 1966 U.S. Naval Institute *Proceedings* article discussed earlier, described many of the fundamental changes brought on by nuclear power. Commander Smith's words, although written more than fifty years ago, still hold true today:

> When the strategist turns from the conventional to the nuclear submarine—and toward that nearing future when navies will face the reality of opposing fleets of advanced nuclear submarines—he moves from a difficult, but finite, problem to one whose very dimensions appear unbounded. For, by the creation of *Nautilus*, the gains of many years of ASW progress were erased. Since then, with the nuclear power plant married to the *Albacore* configuration, the submarine has opened a yawning gap between its own capabilities and those of the ASW forces. Taking departure once more from the fact that the basic virtue of the submarine is its ability to remain hidden, we see in nuclear power an almost infinite multiplication of this capability. When additional assets of high submerged speed (and virtually limitless endurance at that

speed), coupled with the incorporation of the most advanced sensors and weapons of modern technology, are conferred as well, it is manifest that we are witness to something new in naval warships. It is not merely an improved submarine. It is a change of degree so formidable as to constitute a change in kind.

It is only in seeking to discern the shape of a future struggle against such a submarine, however, that we gain full measure of its impact. To begin with, the nuclear submarine virtually nullifies the effectiveness of both the vehicle and the sensor—airplane and radar—that more than any other were responsible for its defeat in World War II. The airplane, deprived of opportunities, will find itself roaming over the surface of an empty ocean, barren of clues, its value narrowing to whatever roles its capabilities will permit in localization and tracking.

And it is the absence of clues as to the submarine's general location that would be one of the distinguishing characteristics of war against the nuclear submarine, and a measure of its increase in difficulty. For it was these clues in World War II, and the uses they served, that were fundamental in defeating the U-boat. These were the clues that made evasive convoy routing effective, led hunter-killer groups in toward their kills, and provided the continuing base of information that enabled offensive ASW forces to achieve suppression and harassment of the submarine from portal to portal. Victory in World War II was a mosaic pattern composed of millions of fragments of incident and encounter, most of them minor in themselves, but together mounting to a high cumulative probability against the submarine's being able to accomplish its mission. It is a pattern that will not exist for the nuclear submarine.[33]

Communications improvements are the second important change which has benefited submarines. By improving how and when they communicate, submarines have greatly reduced two more of their largest World War II vulnerabilities. Today's submariners have learned from the mistakes of Admiral Dönitz and his U-boats to understand the risk every time they transmit.[34] No longer can ASW forces rely on enemy submarines to send frequent, long omnidirectional transmissions which can be detected and plotted using radio-directional finding equipment. Furthermore, it is unlikely that the United States will be able to crack the naval codes of a credible enemy, especially not with the success that the Allies broke the German Enigma or Japanese naval codes in World War II. In the rare instance that an enemy submarine does transmit its location, it will most likely do this discretely and with secure communications. Yet again, submarines have largely corrected one of their major World War II weaknesses, reducing the tools available to enemy ASW forces.

The final major area of submarine improvement is in the form of weapons. While most World War II torpedoes were unreliable unguided weapons launched from exceptionally short range, today's torpedoes are fearsome. Guided by a target's acoustics or homing in on its wake, the torpedoes can be fired at ranges that greatly increase the submarine's striking power. Whereas World War II ships had a chance of evading torpedoes by sighting them early and turning to avoid them, defeating wire-guided weapons with impressive computing power is a much more difficult task.

Submarines are also now capable of carrying ASCMs, further extending their killing range and power. Numerous Chinese submarines can carry ASCMs, such as the Kilo and its 3M-54E Klub (SS-N-27B Sizzler) missiles with a reported range of 220 km.[35] ASCMs also give submarines the capability to fire at targets they have not acquired on any sensor by using tracking information relayed by another source, such as a UAV.

In general terms, submarines' increased speed and weapons range means their effective kill area is orders of magnitude larger than it was in World War II. For example, a submerged submarine operating in 1945 was capable of typical sustained speeds of approximately six knots and could effectively fire torpedoes at ranges of approximately 2,000 yards, meaning that in one hour it could attack any target within an area of approximately 153 square nm. Today, a submerged nuclear submarine, operating at 20 knots over the same time period and using the same torpedo firing range, could attack any target within approximately 1,385 square nm, a 9.1 factor increase in area. If a diesel submarine, operating at 6 knots, fires an ASCM with an approximate range of just half that of the Sizzler, it can still cover an area 88 times that of the World War II submarine. These calculations, although simplistic, show how submarines can threaten so much more of the ocean today. During World War II, warships' best defense was often their speed, as they could sail by a submerged enemy before the submarine could get into range. Today, fast submarines armed with long-range, accurate weapons have time on their side, meaning aircraft carriers and their escorts will be at risk more frequently and for longer periods of time. When those submarines do fire their torpedoes and missiles, they will be high-speed guided munitions with large warheads, not primitive unguided weapons with only mediocre chances of scoring a hit.

Dating back to World War II, when they were vulnerable yet deadly, submarines have undergone important changes that have made them significantly more survivable while being even more lethal to their enemies. The four best methods ASW forces could use to find and attack submarines—airplanes, radar, radio direction finding, and cryptology—are now less credible options. In addition, submarines' unguided, short-range, faulty torpedoes have been replaced by accurate, deadly weapons that attack their targets at high speed above or below the ocean's surface. This portends a future in which aircraft carriers, susceptible to submarine attacks in World War II, will be even more inviting and vulnerable targets in a future war. Rear Adm. William J. Holland discussed this general trend in the *Naval War College Review*:

> Technological advances in sensors, processing, propulsion, quieting, and weapons have made today's submarine a much more formidable opponent to its foes than its ancestors of World War I and II were to their adversaries. Nothing seems to promise to change the relationship; the gap between the submarine and its adversaries will continue to widen. There is no known phenomena which will substantially reduce the submarine's invisibility. The increasing capability of space surveillance coupled with precision navigation, direct communications, and concentrated processing equipments threatens all targets above and on the face of the earth, while aiding those below it.[36]

Opposing those submarines, ASW forces have also made important advances since World War II. The first major improvement in antisubmarine warfare is the new and improved detection methods available, including sonar and magnetic anomaly detection (MAD). Sonar was extensively used throughout

World War II but was often inaccurate and was initially capable of ranges of only approximately a thousand yards.[37] Today, escort vessels have highly advanced passive and active sonar systems with both hull-mounted and towed arrays that are capable of submarine detection at ranges unfathomable during World War II. Maritime patrol aircraft (MPA) and helicopters are now equipped with sonobuoys and dipping sonars, meaning they are no longer restricted to only searching for surfaced submarines. MPA are also often equipped with magnetic anomaly detection, another technology developed during World War II that uses changes in the earth's magnetic field to find submerged submarines. Finally, there are various experimental search methods, such as satellite wake tracking or light detection and ranging (LIDAR) technology.

Despite these advances, each of these technologies have their own problems, some of which reduce their potential to find and track enemy submarines. On the whole, these new methods do not make up for the lost ease with which fleets could find surfaced submarines using aircraft and radar in World War II. Aircraft-based sonar is a powerful tool, but it requires a starting point for an initial search. Short-range helicopters with dipping sonar, and MPA equipped with a limited supply of sonobuoys, cannot be expected to find a submerged submarine in the vast expanse of the ocean. When mounted on surface escorts, any technological advances in sonar have benefited submarines just as much if not more than their adversaries. If the escorts choose to use passive sonar, in which no energy is sent into the water, the submarine has the advantage. Ocean acoustics are extremely variable, but generally speaking the deeper the receiver is positioned and the farther away it is from the noisy ocean surface, the better the reception; this means the submarine is a much better sonar platform. Submarine officers and crews are much better versed in sonar employment than their adversaries because it is the sensor they use every single day while on watch.

Further complicating ASW forces' use of passive sonar is the fact that submarines are much quieter today than they have ever been before. Vice Adm. James Fitzgerald, a former commander of an ASW squadron and a carrier strike group, wrote, "Advances in submarine quieting (loss, or significant reduction, of stable narrow-band acoustic signatures) have largely negated the previous passive acoustics-based long-range detection and reduced the effectiveness of the systems that relied on those vulnerabilities."[38] In his work on underwater acoustics, Dr. Xavier Lurton gathered unclassified estimates of the noise emitted by submarines during World War II and by modern submarines. He estimated that modern nuclear-powered submarines and modern diesel-electric submarines are thirty decibels and forty decibels quieter than World War II submarines, respectively—this equates to one thousand and ten thousand times quieter.[39] Obviously these are extremely rough estimates that are highly dependent on ship class, operating conditions, and environmental factors, but they suffice as a first-order approximation. When considering passive sonar, submarines have the advantage in environment, training, and equipment—meaning they retain the initiative and can decide when and where to attack or break off contact.

If those escorts elect to use active sonar, in which they transmit energy pulses and listen for echoes off submarine hulls, the advantage very well may still rest with the submarine. Active sonar works both ways, in that the searching escort has now broadcast its position to the enemy submarine.[40] That submarine can use this knowledge to home in on a fleet it had not found previously, or to chart the best course for escape now that it knows where the enemy is.

The other new means of submarine detection have their own serious problems. MAD requires a starting point, as its effective range is insufficient to search large swaths of the ocean; in a likely acknowledgment of that challenge, the U.S. Navy's new primary ASW aircraft, the P-8 Poseidon, does not have a MAD sensor.[41] Other technologies—such as satellite wake tracking, LIDAR, and temperature fluctuation measurements—have not proven capable of real-world success. Perhaps these methods would work in laboratory settings or in a constrained area against a target that has already been located, but actual operating environments are much more complicated with numerous variables confusing the problem. There are thousands of objects resembling periscopes, millions of unidentified wakes, and countless unexplained temperature fluctuations in the ocean, and determining which are caused by submarines and which are false positives is a nearly impossible task with the currently available technology. All in all, some improved detection technologies, including sonar and MAD, will aid ASW forces, but these are not as effective as World War II radar-equipped aircraft looking for surfaced submarines.

The second major area of improvement for antisubmarine forces is in weapons technology. World War II aircraft and escorts primarily used unguided depth charges to attack surfaced and submerged submarines. They later gained new weapons like the Hedgehog, which used a mortar-like device to hurl explosives ahead of an escort, and acoustic-homing torpedoes, such as the American Fido.[42] Today, helicopters, MPA, and escorts have their own advanced torpedoes that can target and attack submarines. No longer must ships or aircraft drop dozens of depth charges when they can simply deploy a torpedo in the vicinity of an enemy submarine and let it home in on its target. Furthermore, escorts do not have to sail directly over an enemy submarine thanks to weapons like the U.S. vertical launch antisubmarine rocket (ASROC), which can be fired from cruisers' and destroyers' vertical launch systems, fly ten miles to the submarine's last known location, and then enter the water to commence its attack.[43] However, these new weapons do not solve the problem of submarine detection; they merely help to maintain the status quo from World War II, when submarines were vulnerable if they could be found. Once again, the submarine's most important defense is its stealth, and the easiest way for it to defeat modern torpedoes and weapons like the ASROC is to remain hidden and never give the enemy a target. ASW forces cannot adhere to the tactical maxim of attacking effectively first if they cannot find a submarine to attack.

The most important change in ASW is the use of friendly submarines to hunt the enemy and to protect the surface fleet. With one known exception, submarines did not sink enemy submerged targets in World War II, due in part to the lack of guided torpedoes that could run at variable depths.[44] Friendly submarines armed with guided torpedoes and advanced sonars are now the most effective way to protect high-value units. All the advantages the enemy gains from operating high-quality sonar in the ideal environment can be used against them by a friendly submarine. In addition, when done correctly, the enemy submarine will not know it is being tracked, an improvement over active sonar surface vessel tracking.[45] Adm. Ignatius Galantin recognized the submarine's strength as an ASW platform soon after the launching of USS *Nautilus*; in 1958 he wrote,

> What makes the nuclear submarine of special utility in anti-submarine warfare is the fact that it is the best mobile sonar platform. The importance of this at a time when nuclear power tends to invalidate most other means of detection is self-evident. This improved sonar detection capability comes partly from the ability to operate continuously at the depth best suited to sonar conditions of the day, in a region of low background noise free from the buffeting of surface storms, and partly from the fact that in making submarines as silent as possible there has come the dividend of a well-streamlined, quiet, maneuverable sonar platform that permits taking optimum advantage of modern, long-range sound detection equipment.[46]

Yet there are caveats to this ASW development. Operating against modern, capable enemies, friendly submarines are not a silver bullet that can guarantee the safety of nearby carrier strike groups. As foreign submarines become quieter, it will be even more difficult for friendly submarines to detect, track, and potentially kill these threats before they are able to conduct their own attacks. The idea that U.S. submarines can locate, track, and destroy the dozens of capable submarines operated by countries like China and Russia before they are able to threaten the surface fleet is questionable. Furthermore, even if U.S. submarines effectively stop every submerged threat, there is a large opportunity cost involved. Just as the fleet's cruisers, destroyers, and aircraft sacrifice much of their potential offensive firepower in protecting the aircraft carrier, every submarine used to protect the fleet is an asset not doing what it does best: attacking the enemy fleet. Rear Admiral Holland commented on this problem:

> Here lies the pitfalls within the Navy itself. Submarines have themselves become primary antisubmarine weapon systems. Their presence and performance as part of a task group have built an aura of security and a confidence that, when so assigned, threatening submarines will not appear undetected. This record is admirable but creates a situation that

can dilute the primary task in the event of war. Commanders' demands for submarines to be assigned to protect their task groups subvert the primary attribute of conducting unrestricted warfare against the enemy's forces in waters that otherwise are not open or accessible to others. The proper employment of submarines is as a major force to be wielded as a unit—dispersed and widely distributed under an operational command whose task is to "sweep the seas." Destruction of the enemy fleet is the goal; protecting our own fleet by eliminating the threat is a beneficial byproduct.[47]

Despite important advances for ASW forces, there have not been enough to consistently overcome a submarine's stealth. The winner of an undersea battle often hinges on whether escorts and aircraft can find the submarine, the first and most difficult step in the ASW kill chain.[48] However, as summarized in Figure 6.1, many of the best tools used by ASW forces during World War II are no longer effective today. As the figure shows, the primary means of submarine detection during World War II was aircraft visual and radar searches, surface ship visual, radar, and sonar searches, HF/DF, and cryptology. ASW forces now have fewer, less robust options; the primary submarine detection method today is sonar on aircraft, surface ships, and submarines. No matter how good sonar has become against submerged targets, the basic science of energy transmission in air and water means it will likely never be as powerful a tool as radar was against surfaced targets seventy years ago.

During World War II the Allies expended an enormous amount of time, energy, and resources defeating the German U-boat fleet and winning the Battle of the Atlantic, suffering heavy casualties despite their technological and matériel advantages. In the Pacific, the Japanese devoted much less effort to antisubmarine warfare and lost huge portions of their combatant and merchant fleets as a result.[49] In the more than seventy years since the end of the war, both submarines and their hunters have made important advances, but most of those improvements have benefited the submarine. It has eliminated many of its weaknesses and improved its armament while many of the changes benefiting ASW forces, such as better sonar, also aid submarines. These trends led naval analysts Norman Polmar and Dr. Edward Whitman to conclude their two-volume history of antisubmarine warfare

FIGURE 6.1. Defeating Stealth: Trends in Submarine Detection Methods

Platform or Category	Detection Method	World War II	Today	Trend
Aircraft	Visual	Highly effective against surfaced submarines	Ineffective	Aircraft are less effective at detecting submarines today than during WWII.
	Radar	Highly effective against surfaced submarines	Significantly less effective	
	Sonar	Ineffective	Effective, using sonobuoys and dipping sonars, especially over short-moderate distances	
	Magnetic Anomaly Detection	Mostly ineffective	Somewhat effective over short distances	
Surface Warship	Visual	Highly effective against surfaced submarines	Ineffective	Surface warships are less effective at detecting submarines today than during WWII.
	Radar	Highly effective against surfaced submarines	Significantly less effective	
	Sonar	Somewhat effective, but limited in range and accuracy	Very effective, using powerful new sonars and improved processing capability	
Friendly Submarine	Visual	Somewhat effective	Ineffective	Friendly Submarines are more effective at detecting submarines today than during WWII.
	Radar	Somewhat effective	Ineffective	
	Sonar	Ineffective	Highly effective. Best option for ASW	
Intelligence Methods	Radio Directional Finding	Highly effective, led to numerous submarines' sinking	Significantly less effective	Intelligence is less effective at detecting submarines today than during WWII.
	Cryptology	Highly effective, led to numerous submarines' sinking	Significantly less effective	
	Satellite Tracking	Ineffective	Somewhat effective against surfaced submarines	

by writing, "Overall, in the early 21st Century submarine technologies and capabilities are developing at a significantly faster rate than are U.S.-NATO anti-submarine capabilities."[50]

This study seems to describe a future in which submarines pose a great threat to surface warships. This does not mean that surface warships are suddenly obsolete, but it does suggest that centralizing so much power in one platform whose safety is threatened like never before is inadvisable. With the U.S. Navy's massive size, technological ability, and first-rate officers and sailors, there are few ways it can lose a future war. Having a small enemy submarine sink "the queen of the American fleet, and the centerpiece of the most powerful Navy the world has ever seen," is one way that David could slay Goliath.[51]

SUBMARINE COMBAT SINCE WORLD WAR II

Chapter 4 covered the few instances of conventional air and surface warfare since World War II and highlighted how challenging it is to prepare for a future war with so little recent informative combat data. However, there is even less undersea warfare to learn from since 1945, making the task of predicting submarines' potential in a future war even more difficult. In the decades since World War II there have only been three instances of submarine combat, and even these mostly involved small, obsolete ships. Important lessons can be found in submarine engagements of the Indo-Pakistani War of 1971, the 1982 Falklands Conflict, and the 2010 North Korean sinking of a South Korean corvette; but these isolated, limited conflicts are simply not enough to confirm any theories on the future of undersea warfare.

The first post–World War II submarine combat came on the night of December 2–3, 1971, during the Indo-Pakistani War. The newly commissioned Pakistani submarine PNS *Hangor*, a French-built diesel-electric boat, was operating off the Indian coast when it transmitted a report. Indian forces intercepted the message and dispatched two antisubmarine frigates to the area. These small frigates, INS *Khukri* and INS *Kirpan*, were of the Royal Navy *Blackwood* class. They had a full sonar outfit, antisubmarine mortars, and depth charges, but no ASW helicopters.[52] As the frigates carried out searches for *Hangor*, the submarine stalked them and positioned for an attack. The Indian vessels had not detected *Hangor* when the submarine fired its first torpedo at *Kirpan*, which heard the weapon on sonar and evaded. When *Khukri* advanced on *Hangor*, the submarine fired another torpedo, ripping the ship apart and causing heavy casualties. *Kirpan* conducted an unsuccessful depth charge attack and then left the area at high speed to avoid *Hangor*'s third torpedo.[53]

The *Khukri* sinking demonstrated the difficulty associated with finding a submerged submarine, even if the submarine errs by communicating and drawing ASW forces to its location. The frigates' decision to carry out a slow-speed, constant-course search illustrates an important catch-22 for antisubmarine vessels. When surface vessels operate at slow speeds, they reduce their "self-noise" and thus maximize their probability of detecting a submerged contact. However, operating at slow speeds makes them vulnerable to counterattack and sacrifices their speed, their main defense against a submarine attack. Conversely, it is more difficult for the submarine to follow and attack a high-speed, maneuvering vessel, yet the surface ship may be moving so fast as to essentially blank out its sonar and make it much more difficult to hear anything. Flow-noise, or the sound created by water rushing by the ship's hull, increases exponentially with speed, such that at high speeds there is approximately a two-decibel, or 58 percent, increase in self-noise for each additional knot in speed.[54] What results is another situation benefiting the submarine. If the escort chooses to slow down and conduct an effective search it increases its vulnerability, yet if it operates at high speeds to protect itself its sonar search is seriously downgraded. The Indian frigates chose to conduct a high-quality sonar search yet were still unable to find *Hangor* before it attacked the exposed ships.

After this isolated fight, there was not another instance of submarine combat until more than a decade later, during the Falklands Conflict between Britain and Argentina. The first action came early in the fighting, on April 25, 1982, when a British helicopter used its radar to detect the surfaced Argentinian submarine ARA *Santa Fe*. The submarine, built for the U.S. Navy during World War II, was on the surface after landing reinforcements on the island of South Georgia.[55] *Santa Fe* attempted to sprint to sea on the surface, but after depth charge, torpedo, missile, and machine-gun attacks by four helicopters, the Argentinian crew beached their submarine to avoid sinking.[56] *Santa Fe* confirmed what was already known: surfaced submarines are vulnerable to aircraft equipped with radar.

Several days later the British nuclear-powered submarine HMS *Conqueror* found and started tracking the Argentinian cruiser ARA *General Belgrano* and its two escorting destroyers. The Argentinian vessels were all World War II era ships, but the escorts did carry modernized AN/SQS-29 series sonars.[57] *Conqueror* advanced to within 1,400 yards of *Belgrano* and launched three old Mark 8 torpedoes, two of which hit their target. As the cruiser began to sink, the escorts haphazardly dropped depth charges while the submarine went deep and sped away.[58] As a result of this shocking loss, in which 368 sailors died, the Argentinian navy withdrew its surface fleet to port for the remainder of the war for fear of additional submarine attacks. Throughout the rest of the war, British submarines were stationed near the Argentinian mainland to provide intelligence and advance warning of aircraft attacks.[59]

The only other significant submarine action in the war featured the more modern ARA *San Luis*, a German-built Type 209 diesel-electric boat. *San Luis* conducted a thirty-six-day patrol, but was unable to conduct any successful attacks due to improper wiring of its torpedo fire control panels, causing torpedoes to launch along incorrect bearings.[60] *San Luis* was able to successfully evade all British attempts to find or destroy it, and reportedly operated within the British fleet, allegedly making failed attacks on frigates HMS *Alacrity* and HMS *Arrow*.[61]

The Falklands Conflict featured the most undersea combat since World War II and has important lessons when attempting to understand the modern aircraft carrier's ability to defend itself against submerged threats. However, those lessons are diminished by the fact that the war occurred more than thirty years ago, was short-lived, and involved large numbers of old or obsolete platforms.[62]

The first major undersea warfare lesson from the Falklands is that submarines' stealth gives them a disproportionate effect on the enemy. By not knowing where or how many submarines are present, the enemy is forced to operate as if they are everywhere, and this can have serious effects on both operations and morale.[63] For example, when an Argentinian aircraft overflew HMS *Antrim*, the British canceled an entire operation because they feared their location could be passed to the submarine *Santa Fe*.[64] Additionally, by sinking *Belgrano*, the British convinced the Argentinians that the submarine threat was too great and that they had to withdraw their entire fleet for the remainder of the war. As Vice Adm. Michael Connor, a former U.S. submarine force commander, put it, "Stealth is a force multiplier for the side with undersea dominance, and a paranoia multiplier for the side that does not."[65] Many admirals and public leaders may be rightfully hesitant to risk a carrier strike group in an area known to have lurking submarines, meaning that the enemy can gain at least a partial victory without firing a shot.

The Falklands Conflict also showed how difficult antisubmarine operations can be, especially against modern submarines in the littorals. *Conqueror* operated near *Belgrano* and her escorts with impunity, tracking them for more than a day, penetrating the screen to launch torpedoes from close range, and monitoring the Argentinian rescue efforts after the attack.[66] British ASW efforts were equally futile, as they were never able to find *San Luis* despite the submarine often operating near the British fleet. The *San Luis* captain claimed after the war that the British did not know of his presence until after he fired, and that even then his ship never experienced any counterattacks.[67] Throughout the war ASW forces never located any submerged submarines, yet the British still fired two hundred antisubmarine torpedoes at imagined targets with no effect. As Lawrence Freedman wrote in the *The Official History of the Falklands Campaign*, due to false alarms and submarine anxieties "the Atlantic whale population suffered badly during the course of the campaign."[68]

The U.S. Navy's report on the Falklands Conflict also discussed the difficulty both sides experienced in antisubmarine operations: "The Royal Navy, long believed to be the best equipped and trained Navy in the Free World in the field of shallow water ASW, was unable to successfully localize and destroy the Argentine submarine *San Luis*, known to have been operating in the vicinity of the Task Force for a considerable period. The Falklands experience clearly demonstrates the difficulty associated in shallow waters, an aspect of Naval Warfare which requires increased emphasis in the U.S. Navy."[69] Finally, historian John Keegan concluded, "Consider the record of the only naval campaign fought since 1945, that of the Falklands War of 1982. From it two salient facts stand out: that the surface ship can barely defend itself against high-performance, jet-propelled aircraft; and that it cannot defend itself at all against a nuclear-powered submarine."[70] If the entire British task force could not find a single Argentine submarine in a month, then U.S. efforts to destroy dozens of Chinese or Russian submarines—and thus enable carrier strike group operations—could take years.[71]

The third and final instance of post–World War II submarine combat came when a North Korean minisub sank the South Korean corvette *Cheonan* in

March 2010. *Cheonan* was patrolling in South Korean waters near a disputed maritime border and was employing its active sonar.[72] A single torpedo struck the corvette amidships, causing it to break apart, resulting in the loss of forty-six sailors. An international investigation later determined the torpedo to be a North Korean CHT-02D acoustic/wake homing torpedo, yet little is known of the submarine's motivation or tactics.[73] The submarine evaded South Korean detection efforts and North Korea has repeatedly denied responsibility. The incident did show that "on any given day, any given submarine, no matter how crude or unsophisticated, can sink nearly any surface ship," according to Capt. William Toti, a former submarine captain and fleet antisubmarine warfare commander.[74]

Submarines were flawed but still managed to sink dozens of capital ships during World War II. Since then, technological and tactical advances indicate submarines will be even more capable in the future. Although there is scant modern submarine warfare data to support that hypothesis, in the three available cases of undersea combat since World War II, submarines did prove dominant. In the most striking case, a single submarine—*Conqueror*—launched a single salvo that sank a single ship and in so doing deterred an entire surface fleet from ever leaving port. The historical evidence suggests that submarines will be well poised to deter or sink aircraft carriers, a feat they have already performed fifteen times. The difference today is that the U.S. Navy has centralized almost all its missions and firepower in a small number of those carriers, meaning they hold the key to the fleet's success—or failure—in battle.

A BRIEF HISTORY OF MINE WARFARE

Since their invention by David Bushnell in 1776, naval mines have been an attractive option for both weak and strong navies because of their large potential payoff, their ease of use, and the difficulty associated with defending against them. The first widespread use of mines came during the Civil War, when the Confederacy searched for technologies to help defend its ports and mitigate the large Union naval advantage. Many of their mines, called "torpedoes" at the time, were primitive and became inoperable due to broken moorings, corroded parts, or wet powder. Despite these problems, mines damaged or sank forty-three Union ships throughout the war, making them the Confederacy's most effective naval weapon.[75] For example, during the Union attack on Mobile Bay in 1864, the leading monitor in the Union formation, USS *Tecumseh*, hit a mine and immediately sank.[76] Despite the confusion and risk of additional mines, Rear Adm. David Glasgow Farragut allegedly yelled, "Damn the torpedoes! Full speed ahead!"[77] Farragut's fleet was able to overcome the loss of *Tecumseh*, the threat of additional mines, the powerful ironclad CSS *Tennessee*, and the nearby forts to successfully capture the bay. The city itself did not fall until roughly nine months later, and only after mines had sunk another seven Union ships, including two monitors.[78]

With her impressive armor and firepower, *Tecumseh* was Farragut's most powerful ship, like nuclear-powered aircraft carriers are for today's Navy. *Tecumseh*'s destruction weakened Farragut's fleet, but by no means crippled it. She represented only 25 percent of Farragut's ironclad force, 20 percent of his ironclad cannons, and a mere 1.3 percent of the entire squadron's cannons.[79] In contrast, the most important ship in the fleet today, the aircraft carrier, represents the vast majority of a strike group's firepower. Its loss would result in a drastically weakened force that would be significantly less likely to be able to accomplish the task at hand. Centralizing so much of the fleet's firepower in a single ship is effective when that ship is impervious to enemy attacks, but dangerous when that ship—and all its capabilities—are at risk of being lost all at once.[80]

Half a century later mines were again widely used during World War I. Their most successful strategic use came in the Dardanelles, the long passage linking the Black Sea and Mediterranean Sea. The British and their allies looked to seize control of the strait to allow shipping to flow to and from Russia and to apply pressure on the ostensibly neutral Ottoman Empire.[81] The Turks, with German assistance, had initially laid 343 mines in ten lines and set up mobile and fixed artillery on the shore to protect them.[82] British minesweepers working to clear the area struggled due to the harassing artillery fire, strong currents, and their civilian crews' poor performance.[83] As a result, the combined British and French fleet entered the strait to silence the forts and allow the minesweepers to complete their work. In just a matter of hours, mines aided by artillery fire sank three of the sixteen capital ships and severely damaged another three.[84] After this failed attack, the British admiral leading the Dardanelles operation refused to try again. This decision led to the Gallipoli campaign, a disastrous attempt to take the forts by land that resulted in more than 200,000 British, Australian, and New Zealand casualties.[85]

The Dardanelles showcased the true power of naval mines. They destroyed multiple capital ships, deterred any more from entering the area, and necessitated a massive ground campaign that ended in failure. The campaign also illustrated the impact that defenses and mine replenishment can have on minefield effectiveness. Removing the mines was already a difficult task, but the British minesweepers found it impossible while harassed by the Turkish artillery. In addition, a small Turkish freighter secretly laid an extra twenty mines between engagements which caused much of the damage to the British and French ships.[86] Clearing mines is still a difficult and time-consuming task, one that would be greatly complicated if those minefields were defended. For example, if Iran mined the narrow Strait of Hormuz, it could use its ASCMs—not against the carrier and its Aegis escorts, but against vulnerable minesweepers. Iran could then trust that its mines would keep the American capital ships at bay.[87]

Mines were again extensively employed during World War II; their most effective use was in Operation Starvation. Between March and August 1945, the U.S. Navy and Army Air Force laid more than 12,000 mines around Japan that resulted in 670 sunk or severely damaged ships and contributed to the submarine stranglehold on the Japanese home islands.[88] Unfortunately, after that successful use of mines, the U.S. Navy has often been the victim of these weapons. Over the last seventy years there has been a considerable amount of naval mine warfare—unlike air, surface, and submarine warfare. Of the nineteen U.S. Navy ships sunk or damaged since World War II, fifteen of them have been the victim of mines.[89]

During the Korean War, a planned amphibious landing at Wonsan was significantly delayed because the North Koreans, aided by the Soviets, had laid three thousand mines in the harbor.[90] Minesweeping the harbor required six days and, despite not being under attack by North Korean air or naval forces, four minesweepers and a fleet tug were still sunk and five destroyers were severely damaged.[91] Meanwhile, the 250-ship United Nations task force waited offshore. Its commander, Rear Adm. Allen Smith, famously lamented, "We have lost control of the seas to a nation without a navy, using pre–World War I weapons, laid by vessels that were utilized at the time of the birth of Christ."[92] By the time the Marines got ashore, South Korean troops had already taken the city and Bob Hope was there performing for the U.S. Army.[93]

During the Tanker War in the Persian Gulf, the frigate USS *Samuel B. Roberts* struck an Iranian mine built using a 1908 design. The ship avoided sinking only because of the courageous damage control efforts of the crew. The mine, estimated to cost about $1,500, inflicted $96 million worth of damage.[94] In Operation Desert Storm three years later, the amphibious assault ship USS *Tripoli* hit an Iraqi contact mine that opened a twenty-three-foot hole in the hull; just hours later the cruiser USS *Princeton* was severely damaged by an Iraqi multi-influence bottom mine. The billion-dollar Aegis-equipped cruiser was out of service for the remainder of the war because of a weapon that cost approximately $25,000.[95]

History has shown naval mines to be a simple yet potent option capable of destroying powerful ships and deterring great fleets. New technologies will lead to advances in mine-hunting methods, but the underlying physics of the oceans means it will always be a difficult task. At the same time, mines will benefit from technological advances as well, allowing them to be better hidden and to increase their probability of detonating at the ideal moment. Former Chief of Naval Operations Adm. Frank Kelso acknowledged this shift in 1991, saying, "I believe there are some fundamentals about mine warfare that we should not forget. Once mines are laid, they are quite difficult to get rid of. That is not likely to change. It is probably going to get worse, because mines are going to become more sophisticated."[96]

Mines are effective weapons and deterrents against all ships, not just aircraft carriers. However, the U.S. Navy is not totally reliant on all those other types of ships—it can lose destroyers, submarines, and logistics ships without a fundamental shift in the fleet's combat effectiveness. In contrast, with the Navy relying so heavily on so few aircraft carriers, if mines can deter, damage, or destroy even one, the results would likely be devastating. What happens to a carrier-centric fleet without a carrier at its center?

LESSONS FROM UNDERSEA WARFARE HISTORY

In 1912 a young lieutenant named Chester W. Nimitz wrote a U.S. Naval Institute *Proceedings* article on the value of the submarine. Nimitz wrote, "The steady development of the torpedo together with the gradual improvement in the size, motive power, and speed of submarine craft of the near future will result in a most dangerous offensive weapon, and one which will have a large part in deciding future fleet actions."[97] When Nimitz, promoted to Fleet Admiral, led the Allied victory in the Pacific during World War II, he witnessed submarines achieve numerous victories but pay a steep price for their vulnerabilities. At the end of the war Nimitz predicted, "Battleships are the ships of yesterday, aircraft carriers are the ships of today, but submarines are going to be the ships of tomorrow."[98]

In the seventy years since Nimitz' prediction, there have been significant advances benefiting both the submarine and ASW forces in their struggle, yet the submarine has been the bigger beneficiary. By no longer operating on the surface and using better communications, submarines have eliminated their greatest vulnerabilities, while their drastic improvements in quieting, speed, and weapons allow them to challenge warships like never before. Antisubmarine developments, including improved detection methods and the use of friendly submarines, do not appear to be enough to overcome the undersea advantage. Yet with so little submarine combat to learn from, it is impossible to know how much of a threat submarines will pose to aircraft carriers and other surface ships in the next war. Nothing is more informative than past combat when preparing for the future, but in the last seventy years submarines have only sunk two small escorts and an old cruiser.

Historian John Keegan addressed this problem, and the future of the aircraft carrier and the submarine, while concluding his book *Price of Admiralty*. He wrote,

> Foresight is the riskiest of all means of strategic analysis. It must nevertheless be said that the forecasts of the submariner are intrinsically more convincing. For it with the submarine that the initiative and full freedom of the seas rests. The aircraft carrier, whatever realistic scenario of action is drawn—that of operations in great waters or of amphibious

support close to shore—will be exposed to a wider range of threat than the submarine must face. In a shoreward context it risks attack not only by carrier-borne but also by land-based aircraft, land-based missiles and the submarine itself. . . .

The era of the submarine as the predominant weapon of power at sea must therefore be recognised as having begun.[99]

If he is correct and the "era of the submarine" has begun, the Flex Fleet is a better option than today's carrier-centric fleet. With their poor communications, cargo capacity, and inability to conduct numerous key fleet missions, submarines are not going to replace surface ships; as Capt. Wayne Hughes wrote, submarines are "spoilers."[100] However, the Flex Fleet is designed to better capitalize on their potential while more effectively preventing the enemy from the successful use of the undersea domain.

Against enemy undersea forces, the additional eighty-one *Constellation*-class frigates with their capable sonar suites and ASW helicopters would be able to search vastly more area, better protect high value-units, provide more coverage to prevent the enemy from laying minefields, and free up *Arleigh Burke*–class destroyers for other missions. Shifting many of the carrier's missions to other platforms would allow them to focus on fleet air defense and thus stay farther out to sea, where enemy submarines and mines would be less likely to affect them. Finally, if and when enemy submarines and mines do achieve successes, the Flex Fleet is better structured to absorb those losses and retain its combat effectiveness. On the offensive, with a more distributed fleet of fewer high-value units, more U.S. Navy submarines would be freed from defensive assignments to seek out and attack enemy shipping.

As examined in previous chapters, missiles are both a large opportunity and major threat for the U.S. Navy today. Today's carrier-centric fleet does not capitalize on missiles offensively, maintaining the status quo of relying on carrier aviation, while seeking to generate a perfect defense against them. In contrast, the Flex Fleet seeks to maximize the use of missiles offensively, shift its defense to more practical soft-kill measures, and prepare for when those defenses periodically and inevitably fail. In many ways, these two approaches to the place of missiles in the fleet structure is the same as the approaches to undersea weapons. The carrier-centric fleet demands a near-perfect defense for its invaluable carriers, while the Flex Fleet shifts its focus to attacking effectively first while being prepared for the enemy to inevitably achieve some successes. As regards missiles, torpedoes, and mines, the U.S. Navy can ignore the lessons of naval history by futilely attempting to generate an impenetrable defense, or it can refocus on the offensive, maximizing the use of those same weapons to attack effectively first.

Regardless of the fleet structure, the Navy cannot let the threat of cheap enemy submarines and mines prevent it from deploying its most powerful asset, nor can it risk losing a war if the enemy is able to sink or damage just one large ship. Betting so much on a single, vulnerable platform is to risk the U.S. Navy's large lead in warfighting ability and give the enemy a path to victory it would not otherwise have.

In his 1954 U.S. Naval Institute *Proceedings* article on the future of submarine warfare, Cdr. G. W. Kittredge wrote, "World War II changed all the pre-conceived ideas of submarine warfare. Events, rather than naval staffs, shaped the course of submarine policy."[101] Since then, with the lack of submarine warfare, naval staffs have taken over again. The U.S. Navy should prepare for the day when combat shapes policy by developing a fleet that respects the historical and technological trends in undersea warfare. Those trends indicate that submarines and mines will play dominant roles in future conventional combat, meaning the aircraft carrier will likely never achieve the perfect defense such a priceless asset demands.

CHAPTER 7

UNDERSEA WARFARE TODAY

The upshot is that the seas, at least certain areas of them, are becoming a no-man's-land for surface ships. Whether or not submarines ought to be considered capital ships is beside the point; the carrier will likely not be one.[1]

—Capt. Robert C. Rubel, USN (Ret.)

The previous chapter looked backward from today to understand history's lessons regarding the aircraft carrier's ability to defeat undersea threats and accomplish its missions. This chapter looks forward, reviewing current tactics and technologies to analyze the carrier's survivability and utility. While taking different approaches, the two chapters arrive at the same conclusion: the large nuclear-powered aircraft carrier and its escorts are unlikely to generate the flawless undersea defense needed for a priceless ship, threatening the carrier's ability to execute its varied missions and the U.S. Navy's ability to win wars. The Navy can keep relying on the aircraft carrier while futilely devoting more resources to antisubmarine and anti-mine warfare, or it can evolve its fleet structure to better respect the lessons of naval history and the current status of military affairs.

Modern submarines employ numerous new technologies that make them a significant threat to ships like the aircraft carrier, including nuclear power, air-independent propulsion (AIP), wake-homing torpedoes, supersonic anti-ship cruise missiles, and sonar and fire control systems with impressive computing power. Yet it is submarines' improved quieting that makes them truly fearsome. Passive sonar is by far the most important method to detect submarines, a tool that the U.S. Navy effectively used in the Cold War to track much louder Soviet submarines.[2] However, as submarines' emitted noise levels approach background levels, the basic physics of underwater acoustics mean it will be progressively more difficult to find them. As these reduced acoustic signatures make passive sonar detection of submarines less of a viable option, other scouting methods will help mitigate the problem, but they are all substantially less useful against a capably operated submarine.

Today, more than forty countries have realized the platform's potential; collectively they operate more than four hundred submarines worldwide, a number that continues to grow.[3] Capt. Jerry Hendrix, a retired naval aviator and frequent writer on the aircraft carrier's future, commented on the undersea threat: "Submarines have become the international flavor-of-the-month with regard to nation-state security. Relatively inexpensive export diesel submarine variants from Europe and Australia now provide a credible defensive capability to any country with an ocean shoreline. Torpedoes launched from these boats, and shore- and ship-based missiles can sink outright most of the world's surface combatants and would, at least, significantly degrade the mission effectiveness of American supercarriers."[4]

In this chapter we will first examine the carrier's two most basic options when faced with a submarine threat: fight or flight. By understanding the U.S. Navy's ability to either find and destroy submarines (the fight option) or to keep the carrier strike group mobile and undetected (the flight option), it is possible to predict the carrier's ability to survive and accomplish its mission in the face of a credible undersea threat. We will also briefly review the Chinese submarine fleet as a proxy for any potential enemy. Finally, we will review the other great undersea danger: naval mines.

These reviews are admittedly broad-brush because so many aspects of undersea warfare are open to interpretation and obscured by classification issues. Furthermore, studying equally capable submarines and ASW fleets is not a completely fair examination for today's U.S. Navy, as its enemies' undersea fleets are not of equal quality. Considering the relative advantages of modern submarines and ASW forces can nonetheless shine a light on the aircraft carrier's vulnerability to undersea threats, prepare the U.S. Navy for a future when its enemies do have more capable submarines, and provide clues as to what type of fleet the U.S. Navy should be building.

Together, those reviews provide strong evidence that submarines and mines pose a great threat to the carrier strike group; indeed, Captain Hendrix expressed his opinion that the carrier strike group is only effective when operating in a permissive environment. He also warned that such a crucial asset should only be assigned to operate in an anti-access environment "under the most extreme conditions when national interests compel leadership to risk" the loss of such a huge investment in hardware, personnel, and prestige.[5]

THE FIGHT OPTION: WHO CAN "ATTACK EFFECTIVELY FIRST"

Capt. Wayne Hughes focused much of his seminal book, *Fleet Tactics and Coastal Combat*, on the importance of attacking effectively first. He wrote that this "great naval maxim of tactics, Attack effectively first, should be thought of as more than the principle of the offensive; it should be considered the very essence of tactical action for success in naval combat." He added, "To attack effectively (by means of superior concentration) and to do so first (with longer-range weapons, an advantage of maneuver, or shrewd timing based on good scouting) have been the warp and woof of all naval tactics. Everything else—movement, cover and deceptions, plans, and [command, control, and communications]—has been aimed at achieving such an attack."[6] Borrowing from Captain Hughes, this section will attempt to determine which—the submarine or the carrier and its protectors—is more likely to be able to "attack effectively first."

This duel can perhaps be reduced to a contest of which side can first find the enemy and then effectively launch weapons at it. In general terms, the submarine has poor scouting ability and poor weapons range but is extremely difficult to find. Conversely, the surface fleet has excellent scouting ability and weapons range, but is comparatively much easier to find. Scouting ability and weapons range combine to define the likely victor in any struggle. The submarine cannot win if it can locate the carrier, only to find that it is not within the range of its weapons. Similarly, the carrier cannot win if it can hit targets at incredible ranges but cannot find the enemy submarine within those ranges.

A study of scouting ability and weapons range may provide insight into which force can attack effectively first. As Captain Hughes wrote, "Nothing about naval combat is understood if its two-sided nature is not grasped. Each side is simultaneously stalking the other. Weapon range is relative to the enemy's weapon range. The weapon range that matters is the productive range—that is, the range at which a telling number of weapons may be expected to hit their targets. The weapon range that matters, for a battle force, is the range at which enough weapons can be aimed to hit with great effectiveness."[7]

For these purposes, scouting encompasses all means of searching, detecting, tracking, and targeting enemy forces.[8] Determining which side has the relative advantage in scouting is, at its most basic, a competition between the submarine using its poor vision to find something in plain sight, and the surface fleet using its excellent vision to find something in hiding. By examining the general means of scouting—to include sonar, magnetic anomaly detection, radar, electronic support measures, visual, and external cueing—it is possible to predict which side is more likely to locate the other one first.

Sonar is by far the most important scouting method for undersea warfare. Passive sonar involves the use of sensors to detect noise radiating from a target, whereas active sonar uses a signal reflected off the target that is then analyzed by the platform that originally transmitted it, in much the same way that radar functions. The noise emanating from targets and detected using passive sonars can either be broadband or narrowband. Broadband noise is spread out over a wide range of frequencies, whereas narrowband is centered at a specific frequency. Broadband noise can be caused by numerous factors, but is typically dominated by propeller noise and cavitation, machine noise, and flow noise (which is caused by the turbulent movement of water over the ship's hull). Narrowband noise is generated by specific machines, such as rotating pumps, reduction gears, and engines.[9]

A ship's ability to detect either broadband or narrowband noise is governed by the strength of the signal generated at the source, losses in the water between the source and receiver, the background noise present, and the ability of the receiver's sonar array to analyze signals while filtering out the background.[10] Because broadband signals often resemble background noise they can be difficult to detect at long ranges where their reduced amplitude is similar to that of the surrounding ocean.[11] However, it is possible to detect narrowband signals at much greater distances because the energy is focused on one frequency instead of being spread out like in broadband noise. In addition, narrowband sources like pumps and gears tend to be low frequency, which do not suffer as many transmission losses and thus propagate farther.[12] However, underwater acoustics are highly variable and notoriously difficult to analyze, and are affected by water depth, currents, surface weather, location of the source and receiver, and a multitude of other factors. The study of underwater acoustics is a science, but its interpretation is often an art.

Despite this variability, when considering submarines pitted against surface escorts, submarines hold the advantage in both broadband and narrowband passive sonar. This advantage stems from each platform's emitted noise levels, how that noise propagates through the ocean, and the detection and analysis of those signals.

Submarines' first advantage over surface escorts is their emitted noise levels. As discussed, a primary component of emitted broadband signature is the ship's flow noise as water rushes over the hull. This noise intensity increases exponentially with speed, such that at high speeds there is approximately a 1.5- to 2-decibel increase in self noise for each additional knot in speed.[13] Because surface ships tend to operate at higher speeds than submarines, this puts them at a disadvantage. As a result, with all other factors being equal, even a five-knot difference in speed means that the surface ship emits ten times as much flow noise.

Another primary source of broadband noise is propeller cavitation. Cavitation is the process of formation of vapor bubbles in low pressure areas, such as on propeller blades, and their subsequent collapse as they enter higher pressure areas. Cavitation is a major source of noise in the oceans and is affected by the propeller design and rotational speed as well as the pressure of the surrounding water.[14] Submarines are significantly less likely to produce cavitation because they usually operate at slower speeds and more importantly, because they operate at deeper depths where the increased pressure makes bubble formation less likely.[15]

The final primary component of a vessel's broadband signature is its general machine and transient noise. Submarines also hold an advantage here because they are designed, constructed, and operated with stealth as a singular focus. In 1958, Adm. Ignatius Galantin discussed this focus, writing that a submarine's concealment "is the quality which must underlie all proper applications of submarine warfare."[16] Even louder than the escorts are the aircraft carriers themselves. As former Deputy Undersecretary of the Navy Seth Cropsey (a retired surface warfare officer), Cdr. Bryan McGrath, and defense consultant Timothy Walton wrote in their report on the aircraft carrier, titled "Sharpening the Spear," the carrier "has a relatively large acoustic signature, generating noise from aircraft takeoffs and landings, hard-mounted propulsion and power generating machinery, and low propeller speed onset of cavitation." As a result of all this noise, the Defense Science Board reported that, under certain conditions, carriers can be detected at ranges of several hundred miles.[17]

In addition to emitting less broadband noise, submarines have the edge in emitted noise levels because they have quieted many of their narrowband sources. Dr. Owen Cote Jr., associate director of the Massachusetts Institute of Technology (MIT) Security Studies Program and a former researcher at the Center for Naval Analyses, wrote, "The enemy of passive, low frequency, narrowband signal processing are [sic] quieting techniques that insulate a submarine's rotating machinery from its hull, preventing the coupling of machinery vibrations to the hull and into the surrounding water, and thereby reducing the acoustic source level of the submarine's tonals below the background noise at those frequencies." These narrowband tonals made detection of Soviet submarines possible, yet today's designs and construction are advancing to a point where narrowband tracking is more and more difficult. Vice Adm. James Fitzgerald, a former commander of an ASW squadron and of a carrier strike group, wrote, "Advances in submarine quieting (loss, or significant reduction, of stable narrow-band acoustic signatures) have largely negated the previous passive acoustics-based long-range detection and reduced the effectiveness of the systems that relied on those vulnerabilities."[18]

Once the broadband and narrowband noise is generated at the source, it must pass through the water to the sonar receiver. Transmissions received at submarines tend to be stronger than those reaching surface ships due to their different operating environments. Due to density's effects on underwater sound, it is generally preferable to have a deep rather than shallow receiver. Submarines can analyze their surroundings and adjust their depth to maximize their detection ability, whereas ASW escorts are confined to the surface.[19] Furthermore, there are often thermal layers in the ocean which greatly affect sound propagation and make it very difficult if not impossible for escorts to detect submarines below. In his book on Chinese submarines, Dr. Peter Howarth wrote, "Submarine commanders are able to exploit the salinity and temperature layers of the water, the acoustic properties of the environment and the topography of the seafloor to manoeuvre tactically in the same way that commanders on the land exploit the physical terrain."[20]

When considering passive sonar, the final reason submarines have the edge over surface escorts is due to the reception and analysis of those signals. Just as it is difficult to hear quiet noises in a loud room, the surrounding background noise plays an important role in interpreting sonar. Surface ships have substantially more flow noise and must deal with more noise from waves, wind, and nearby shipping, all problems that raise their background levels and make it more difficult to detect quiet submarines.[21]

Once these signals are detected, the submarine likely has the advantage in sonar's analysis thanks to the focus and experience of its officers and crew. The average surface officer or sailor is likely proficient at antisubmarine warfare, but he or she must also be an expert in areas including antiair warfare, ballistic missile defense, amphibious warfare, fleet maneuvers, underway replenishment, helicopter operations, and drug interdiction.[22] This comes at a cost, and it means the average submariner—who knows essentially nothing about these missions—can become a true expert on sonar and its use in anti-surface warfare.[23] In addition, the average submarine officer has extensive experience tracking hundreds of surface ships at sea, practicing it every day while on watch. On the other hand, the average surface warfare officer has considerably less experience tracking submarines purely as a function of how few of them there are available for training exercises.

When considering passive sonar—the most important tool for undersea warfare scouting—the submarine clearly has the advantage over an equally capable surface escort. It is built and operated to emit less noise, it operates in the ideal environment to maximize its detection capability while minimizing its chances of being heard, and it has important advantages in signal detection and analysis. Cdr. Robert Smith acknowledged this advantage in his 1966 U.S. Naval Institute *Proceedings* General Prize Essay Contest winner:

> Sooner or later, amidst concentration on problems the submarine presents, doubts are likely to take shape and loom in the background of one's thoughts. Doubts, for instance, that it is reasonable to expect that a surface ship, existing in the turbulent, interface of air and sea, inherently noisier and visually detectable, can ever attain a capability to match that of a submarine in which multiple virtues of invisibility, adaption to a single environment of limited variability, and a concentrated focus of mission are harmoniously joined. And the doubts, all coalescing, add up to the fundamental question of whether, weighing the clear and demonstrable evidence of the submarine's great, and still growing, capabilities against the uncertain gains of ASW, we are not possibly witness to a historical trend which will culminate in the ascendancy of the submarine as the decisive arbiter of naval power.[24]

To better understand submarine versus surface ship passive sonar, imagine two people searching for each other in a large field during a dark night. One person has a significant advantage because he is not only quieter than his foe, but is a better listener, too.

If surface escorts elect to use active sonar, they will find that it can certainly help them find enemy submarines, but is far from a perfect solution. When surface ships use active sonar, they transmit a pulse of energy into the surrounding water that reflects off targets and can then be detected and analyzed by the ship. When used correctly, it can yield high-accuracy firing solutions on submerged threats. However, its limited range affects its utility and it can often backfire on the transmitting escort by alerting the enemy.

Typical active sonars are in the high and medium frequency range in order to achieve the requisite target resolution and to keep the transmitting arrays to a reasonable size.[25] However, these frequencies experience high levels of attenuation and reverberation in the water, causing their power to dissipate quickly. Consequently, the range at which active sonars can accurately detect targets is often limited. MIT's Dr. Cote suggests that "extremely powerful active sonars have difficulty achieving detection ranges of more than thirty miles," and Adm. James Stavridis, commenting on his time commanding a destroyer, wrote that active sonar has a range of about twenty-five miles.[26] This range is cut even further in the littorals, where the shallow bottom, the water's surface, and the large number of biologics create numerous echoes and rapidly attenuate the pulse's energy, making it very difficult to find lurking submarines.[27]

Additionally, escorts' use of active sonar benefits enemy submarines. Unlike passive sonar, active sonar puts energy into the water and alerts the target that it is being hunted. That energy can be used by the hunted to determine the location of the ASW hunter. The active pulses suffer transmission losses over two directions for the surface ship (from the ship to the submarine and then back) versus just one direction for the submarine (from the transmitting ship to the submarine). This means that the submarine can detect the active pulses from the surface ship at a range that is double that of an equally capable surface ship searching for a submarine.[28] Even if the submarine is well outside the range for accurate solution development, the active pulse could still be strong enough to at least alert the submarine of the presence and direction to a transmitting escort. For example, even in the early 1970s U.S. submarines participating in exercises were tracking SQS-26 active emissions at ranges of 150 to 225 nm, whereas the transmitting destroyers were only able to detect the submarines at ranges of approximately 30 nm.[29]

This presents a catch-22 for the surface fleet: active sonar is one of their best tools for finding enemy submarines, but it can also attract threats. Surface ship active sonar is very effective for short- and medium-range searches and to clear the immediate vicinity of submerged threats, but comes with serious drawbacks and is a large liability over medium to long distances. Again, imagine two people searching for each other in a large field during a dark night. If one person switches on a flashlight, he may be able to effectively search the immediate vicinity, but he has given away his own location in the process.

In addition to surface escorts, antisubmarine aircraft can also employ sonar to protect the aircraft carrier. These aircraft, including helicopters and MPA, can use sonar in ways unavailable to surface platforms that in many respects make them better submarine hunters. Their best attribute is speed, which allows them to quickly deploy to an area where a submarine has been detected and track it before it disappears. ASW aircraft can use their speed and range to track submarines long before the threat approaches the fleet; they also have their own weapons to launch attacks. ASW aircraft can accomplish all these missions without fear of being attacked by the submarine.

On the negative side of the ledger, ASW aircraft have several serious limitations that make their task of finding and destroying capable submarines challenging. If surface escorts struggle to find capable submarines with passive sonar, then ASW aircraft—with smaller arrays, less computing power, and less endurance—likely have an even more difficult time detecting submerged threats. If large surface escort active sonars, powered by massive diesel engines and gas turbines, can only achieve ranges of approximately twenty-five miles, then dipping sonars and sonobuoys must be significantly less capable simply due to size, weight, and power restrictions. In addition, ASW helicopters, flying from the carrier strike group, may not be able to use active sonar for fear of giving away the fleet's position. Basic logistics also negatively affect the sonar search capability of ASW aircraft. These aircraft can only carry a limited supply of sonobuoys and their flight times mean they typically cannot provide continuous coverage. Finally, these platforms—helicopters that must hover to deploy their sonars and large, lightly defended MPA—are some of the most vulnerable military aircraft in the skies. Their ability to conduct their ASW mission in contested airspace against a credible enemy is questionable, forcing them to stay close to the aircraft carrier and again limiting their effective search range.[30]

All these factors seem to indicate that ASW aircraft sonar reinforces the carrier's short- and medium-range defenses while having limited capacity for medium- and long-range searches. In many ways, their equipment and capabilities are tailored to patrol the immediate vicinity and to clear selected areas, but not to search for an undetected submarine. These disadvantages are compounded by the improved quieting technology incorporated into modern submarines. As Lt. Cdr. Ryan Lilley, a P-3 pilot, wrote, "With respect to submarine stealth, a modern, quiet submarine holds most of the advantages. . . . The high source levels that NATO navies used to exploit are simply not there anymore. ASW assets like the P-3 have seen their effectiveness decline as they are forced to reduce their search-area size in order to have a reasonable probability of detection."[31]

The best sonar platform for ASW work is a friendly submarine. Friendly submarines negate all of the enemy's advantages over surface escorts by also using the best sonar equipment in the ideal operating environment with the most experienced officers and crews. Submarines can track their foes without announcing their own presence, making their enemies unaware that they are being followed and about to be destroyed.[32] These advantages in equipment, conditions, and personnel make submarines by far the best ASW platform available.

Although submarines do not have many of the inherent disadvantages of other platforms for ASW work, they have their own serious issues that significantly hinder their ability to protect a carrier strike group. Many of these issues stem from the fact that submarines are outstanding platforms for offensive operations, but their poor communications and scouting ability make them significantly less capable defensively. The carrier strike group's best defense against submarines is its mobility and speed. However, a friendly submarine must either operate at very high speeds to keep up, and thus drastically reduce its search capability, or drop far behind in order to conduct an effective search. As the carrier strike group changes plans and courses, it must frequently communicate with the submarine, a challenge that reduces the submarine's search capabilities.[33] Retired submariner Rear Adm. William Holland discussed some of the issues affecting the use of submarines to defend surface fleets, writing, "Since their invention submarines have rarely been incorporated effectively in Fleet movements and operations. Because of their limitations, they have been adjuncts at best. Even when deployed in support of Fleet operations, their roles usually have been as independent operators."[34]

The biggest issue facing friendly submarines in ASW is that they, too, can struggle to find quiet, properly operated enemy threats. There is a limit to their passive sonar's ability to detect targets, and as capable submarines get progressively quieter, the range at which they can be detected becomes commensurately shorter. Capt. James Patton, a retired submarine captain and frequent writer on undersea warfare, pointed out that U.S. submarines had a roughly forty-decibel advantage over their Soviet adversaries during the Cold War, meaning that Soviet submarines were approximately ten thousand times louder. He went on to say that U.S. submarines no longer enjoy that massive advantage over improving enemies, and even when there is an acoustic advantage, "it can sometimes equate into initial detection ranges and detection advantages measured in hundreds of yards rather than many 10s of miles. The bad news associated with this reality is that it is often not a viable option for even a modern, quiet submarine to sanitize a given geographic area looking for other modern, quiet submarines, since the mathematically derivable *mean time to detect* is unacceptably large."[35] There is only so much signal processing sonars can accomplish, and as a submarines' emitted noise approaches background sound levels, those sonars will only be able to detect enemy threats at decreasing ranges. Again, imagine two people quietly searching for each other in a large field during a dark night. This time, both people silently wander around the field and if they ever do find each other, it takes a great deal of time to do so.

Sonar is the most important scouting method in undersea warfare. The factors discussed in this general analysis indicate that submarines have a significant advantage in sonar over ASW forces. Submarines are a platform designed, constructed, and operated to maximize their sonar detection capabilities while minimizing their own acoustic vulnerability, allowing them to be more likely to find the enemy than be found themselves. Retired submarine officer and Naval War College professor William Murray analyzed the submarines' inherent advantage:

> Properly operated, modern, quiet submarines are very difficult (for even the most sophisticated and well-trained opponents) to detect and attack. Simultaneously, such submarines, equipped with modern, easy-to-use weapons, are a potentially deadly nemesis for surface naval forces, which are significantly easier for submarines to detect and attack. This imbalance arises not from misplaced training priorities, unrealistic exercises, inadequate funding, or other forms of bureaucratic neglect (though all those causes can exacerbate the problem), but instead, from physics. Physical laws and limitations, akin to gravity and just as difficult to overcome, are what make antisubmarine warfare exceptionally difficult.[36]

While sonar is a relative advantage for submarines, there are other scouting tools to consider. Two of the better options for ASW helicopters and MPA are MAD systems and radar. MAD systems are designed to find submarines by detecting their distortion of the Earth's magnetic field, whereas radar is used to detect submarine periscopes or other masts. However, physics again acts against ASW forces in these cases. Magnetic field strength decreases rapidly with distance, meaning that MAD systems are only effective at relatively short ranges, and are further hampered if the aircraft is too high or if the submarine is too deep. Furthermore, there are countless metallic ores and sunken wrecks in the ocean which can create false positives. MAD is likely effective against already located submarines, but like many other ASW tools is not suitable for wide-area search. In addition, the radar cross section on periscopes and other masts is obviously miniscule and is comparable to the innumerable objects floating in the sea. Finding submarines using radar also requires them to be cooperative and keep masts out of the water for extended periods of time, unlikely in an era where nuclear and AIP submarines rarely if ever need to come up for air. Even conventional submarines, when operated properly, are unlikely to expose themselves by snorkeling with air threats radiating nearby.[37]

Electronic support measures (ESM) involve the use of detected energy, such as radar, to locate a transmitting enemy. Considering this tool, the advantage is wholly in the submarines' favor. Properly operated submarines rarely if ever transmit any energy, including communications and radar, as they are cognizant of how it can be used against them.[38] Conversely, carrier strike groups often transmit a multitude of navigation, surface-search, and air-search radars. With distinctive sources like the Aegis SPY-1 radar, the dual-band radar on *Ford*-class carriers, and numerous communications circuits, it is not overly difficult to identify the source of those varied signals. All of that energy helps the fleet find and attack aerial and surface threats, while also providing a beacon that enemy submarines can home in on. As naval analyst Norman Polmar wrote, "The CVNs are large, conduct highly detectable air operations, invariably radiate electronic systems, and their four screws produce unique acoustic signatures."[39]

Visual methods of detection boil down to the submarine using one person to search for several huge targets, and the surface and air fleets using many people to search for a tiny target. Yet again, at extended ranges the submarine likely has the advantage, although at shorter ranges ASW forces may be able to detect the submarine's periscope depending on time of day, visibility, sea state, and the submarine operators' skill. Platforms with infrared sensors may be able to detect exposed masts at night, but this again requires a cooperative target.

The final category of scouting methods can be referred to as "external cueing"; it includes any platforms or sensors that operate separately and then transmit information to the carrier strike group or submarine. In this category, the submarine is at a significant advantage at all ranges. Whether targeting information is provided by satellites, passing merchant ships, UAVs, or land-based coast watchers and radars, the submarine is virtually invisible to all these options while the carrier strike group is substantially more vulnerable. All of these tools give submarines another means of finding their surface foes without exposing themselves in the process. Rear Admiral Holland discussed this imbalance, writing, "There is no known phenomena which will substantially reduce the submarine's invisibility. The increasing capability of space surveillance coupled with precision navigation, direct communications, and concentrated processing equipments threatens all targets above and on the face of the earth, while aiding those below it."[40]

This section presents an overview of the primary scouting methods in undersea warfare and reveals which side has the advantage in each. We will only study the primary ASW options; for example, ocean surveillance ships equipped with the surveillance towed array sensor system (SURTASS) will not be discussed because of how vulnerable these large, unarmed vessels would likely be in a time of war.[41] However, this approach is far from perfect in part because undersea warfare is not one platform using one sensor against the submarine at one time. Rather, it is a "team sport" where the fleet's air, surface, and submerged platforms are simultaneously using every available tool to try to detect, locate, and destroy the enemy threat.[42] By getting intermittent whiffs of that submerged threat, ASW forces attempt to refine the enemy's location enough to focus their search efforts on only a limited area.

Yet despite this cooperative approach, the evidence still favors submarines in terms of relative scouting ability. As submarines get quieter and more capable, they will give fewer and fewer clues to their adversaries. This will force ASW commanders to spread out their limited resources on larger swaths of the ocean, giving the submarine more time to locate and attack the surface fleet. That fleet—visually obvious, acoustically loud, and radiating a great deal of trackable energy—will be difficult to keep hidden long enough for its protectors to find and kill the enemy submarine. As Captain Patton wrote, "Submarines operating covertly at slow speeds in acoustically difficult waters represent a Herculean search and localization effort for even the very best ASW platforms, including modern SSNs, and if not essentially impossible, involve a statistical length of search that is unacceptable if time is of the essence (which it always is) to degrade the [anti-access/area denial] effort."[43] Adm. James Stavridis also discussed the difficulty associated with ASW in his book *Destroyer Captain: Lessons of a First Command*, writing, "Destroying a submarine, I think, is the hardest task in naval warfare."[44] This analysis of relative scouting ability is far from conclusive. It seems that carrier strike groups are effective at submarine scouting in the short range, but that submarines are much more likely to find the surface fleet before being found.

The submarine's ability to find the carrier strike group before being found itself gives it a massive tactical advantage. In the Age of Sail, tactical positioning was often decided by which ships held the best position as determined by the wind, sun, and nearby land. Today, the force that locates the enemy first, such as at the Battle of Coral Sea, can attack without being attacked. As Italian naval theorist Giuseppe Fioravanzo wrote, "The fundamental tactical position is no longer defined by the *geometric* relationship of the opposing formations, but by an *operational* element: the early detection of the enemy."[45]

In addition to scouting ability, the other half of determining who can "attack effectively first" is weapons range. Here, the carrier strike group has an undisputed advantage over the submarine. Its helicopters and supporting MPA can fly hundreds of miles to drop their torpedoes, surface escorts can sail independently and then launch antisubmarine rockets far from the fleet's location, and friendly submarines can attack the enemy with their own torpedoes. If the enemy submarine is found, the fleet has numerous long-range platforms and weapons it can use to attack. However, this is a big "if."

Combined, scouting effectiveness and weapons range yield Captain Hughes' "effective weapons range." This study shows that in broad terms, this submarine versus carrier struggle is decided by each platform's most limiting step: the weapons range for the submarine and the scouting range for the carrier. Considering these limiting steps, the ranges for submarines primary weapons—torpedoes and missiles—are both greater than the carrier's detection range.

When considering the submarine-launched torpedo, the submarine appears to have a slight advantage, but it comes at a considerable risk. In order to effectively use the torpedo, the submarine must close to a range that is somewhat comparable to the range at which carrier strike groups can be expected to detect submerged threats. In this situation, the victor may be decided by differences in ship quality, the skill and training of the men and women on each side, and luck. Here, the submarine's element of surprise tips the scales in its favor. It can prepare for the attack and focus all of its energies for a short period of time, whereas the ASW defenders must be constantly vigilant. As Vice Admiral Fitzgerald described ASW, "It is '24/7,' monotonous and boring punctuated by periods of intensive operational uncertainty."[46] All things considered, modern capable submarines seem to have a good chance of putting torpedoes into defended capital ships.

The submarines' other offensive option—ASCMs—greatly increases its effective weapons range and tips the scales even further. Instead of having to expose themselves by sailing toward the carrier's credible ASW defenses, ASCM-equipped submarines can remain in the shadows and use their own sensors or external cueing to launch missile attacks. In the struggle of submarine weapons range against ASW detection ranges, ASCMs completely shift the balance in favor of submarines. They force ASW forces to patrol a much larger ocean area in order to ensure the aircraft carrier's safety, stretching their resources even further. For example, if a submarine torpedo has a range of an assumed 15 miles, then the carrier strike group must effectively patrol an area of approximately 700 square miles to ensure that no submarines are in firing range. However, if that submarine is armed with ASCMs with a range of fifty miles, then those same ASW forces must effectively patrol an area of approximately 7,800 square miles, a daunting tenfold increase.

This section endeavored to determine which side has the greater "effective weapons range." Determining who can both find the enemy and then be in range to attack it can help determine who is able to accomplish Captain Hughes' tactical maxim of being able to "attack effectively first."[47] This first-order analysis of relative scouting ability and weapons range seems to indicate that submarines have a moderate advantage when considering torpedoes and an even greater advantage if they employ ASCMs.

To summarize this idea, yet again imagine two people searching for each other in a dark field. One has a flashlight and can effectively survey the area around him, while the other lurks in the shadows with a knife. That attacker can use guile and surprise to get close enough to attack his foe, but he also puts himself at risk when he enters the flashlight's range. But if the attacker is armed with a slingshot, he can safely stay in the shadows far from his enemy while launching deadly attacks.

THE FLIGHT OPTION: CAN THE CARRIER AVOID THE SUBMARINE?

The carrier strike group can triumph over the submarine by finding and destroying it, but it can also accomplish its mission by simply avoiding the submerged threat. The surface fleet's mobility and speed are its best defense against submarines. If the submarine can never find or close with its target, the carrier strike group can go about accomplishing its mission without ever directly confronting the undersea threat. In World War II this mobility protected capital ships from countless submarines, who watched their prey sail over the horizon without the means to catch up. When submarines were within range of capital ships, it was often a result of luck; the surface fleet just happened to unknowingly sail in the vicinity of the enemy submarine and it took advantage of the situation.

The mobility of the carrier strike groups is still an excellent defense, but three important technological changes reduce its effectiveness. As submarines get better at finding surface ships, employ improved speed, and use highly capable weapons that greatly expand the threat area, it will become progressively more difficult for surface ships to avoid credible submarine attacks.

The first factor mitigating the carrier strike groups' mobility advantage is the improved tracking and cueing methods available today. As submarine and external sensors, including satellites, become more advanced and numerous, there will be more and more clues that submarines can use to narrow their search areas. When the carrier and its escorts are found, even if it is only briefly, it provides a starting place for an enemy submarine to close in on its prey. Norman Polmar discussed how modern technology makes it "impossible to hide" an aircraft carrier at sea. He wrote,

> Aircraft carriers have long been dependent on mobility for their survival, with their mobility making it difficult or impossible to pre-target them. Carriers are now vulnerable to continuous tracking by satellite and long-endurance unmanned aerial vehicles (UAVs). Related to this situation, modern carrier operations cannot be conducted in an electronically quiet or "emission control" environment, which, coupled with the energy that a nuclear carrier puts into the water when she moves, makes her impossible to hide.[48]

The second major factor eroding carriers' mobility defense is the increased speed and endurance of submarines. In order to attack defended warships during World War II, submarines generally needed to be submerged to avoid enemy escorts and aircraft. However, when submerged, they were limited to a top speed of roughly nine knots for short periods of time.[49] However, today's nuclear submarines can sail at substantially faster speeds indefinitely, meaning they can quickly reposition when cued in on an enemy fleet, and have the potential to catch up to that fleet once found. While conventional diesel-electric submarines generally do not have this option, AIP submarines—able to sail submerged for weeks before snorkeling—have been described as a "poor man's nuclear submarine." Professor Murray explained how AIP submarines could "sprint at high speed to intercept and attack with a wake-homing torpedo a carrier battle group transiting at high speed to a war zone," and then use their AIP system to creep away without needing to expose themselves by snorkeling.[50]

The final issue affecting carrier mobility is the speed and range of submarine weapons. No longer do submarines need to approach within two thousand yards of their target to fire their weapons. Wake-homing or acoustically guided torpedoes with impressive ranges greatly expand the area a submarine can effectively threaten. That area is expanded even further by ASCMs. These weapons, with their own guidance systems, mean that submarines have more time to prepare and launch their weapons than ever before.

Captain Hughes identified this shift in the importance of maneuver. He wrote that there is a trend in maneuver today shifting the emphasis from "speed of platform to speed of weapon. Until World War II maneuvering the fleet was the very heart of tactics. During the war aircraft speed took precedence over ship speed. Since the war missile speed and range have created a tactical environment in which weapons will be delivered without much change in ship position."[51] When five British ships chased *Constitution* in the War of 1812, the few extra knots of speed she got from kedging (running an anchor ahead of the ship, dropping it, and then hauling in to move the ship forward) decided the encounter and let *Constitution* escape her foes.[52] Today, the carrier's ability to sail ten knots faster than many of its attackers is still an advantage, but that advantage means a lot less when considering weapons moving at hundreds of knots over hundreds of miles.

Even if the carrier and its escorts are able to periodically avoid the undersea threat, the submarine still has the advantages of time and probability. That submarine only needs to successfully find and attack the carrier once to accomplish its mission, whereas the carrier must use its mobility to evade the submarine each and every day in order to win. The longer a conflict lasts, the smaller the probability that the carrier can continue to successfully evade enemy submarines.

Thus far this chapter has taken a simplified approach to the prediction of the modern submarines' ability to threaten aircraft carriers. The evidence suggests that both of the carrier's antisubmarine options—fight or flight—are less viable choices today. Submarines have improved scouting and weapons ranges, whereas the carrier's ability to find undersea threats is worsening as submarines become increasingly quieter. These trends indicate that submarines are gaining the upper hand in the ability to "attack effectively first," and that the carrier's mobility is not the defense it once was.

This analysis does not suggest that surface ships are obsolete or that submarines have anywhere close to an absolute advantage. It *does* mean that centralizing the fleet's firepower on one loud, visually obvious, constantly radiating vessel, no matter its offensive capabilities, is far from the best course of action. To be best prepared for any future war, the U.S. Navy should acknowledge the trends in undersea warfare, accept that submarines are very likely to be highly effective against large surface warships, and design a fleet that can still accomplish its mission even after enemy submarines have had their say.

PLAN SUBMARINES

Up to this point in the chapter, the undersea warfare analysis has been a discussion of capable submarines and equally capable ASW forces. However, this hypothetical comparison is not an accurate representation of the actual submarine threats facing the U.S. Navy's aircraft carriers today. The submarines of many of America's competitors are highly capable and improving, but most experts agree that the U.S. Navy ASW and submarine forces have a significant advantage over their foes.[53] As former Chief of Naval Operations Adm. Jonathan Greenert said, the undersea domain "is the one domain in which the United States has clear maritime superiority."[54] To transition from hypothetical to actual submerged threats, this section studies the Chinese submarine fleet and its ability to threaten U.S. aircraft carriers. With the largest fleet in the world, the People's Liberation Army Navy (PLAN) offers a suitable proxy for any potential adversary the U.S. Navy may face.

For much of its early existence, the PLAN submarine fleet was obsolete, loud, and poorly armed. In the 1990s three-quarters of Chinese submarines were locally built variants of the Soviet *Romeo* class, and the few nuclear-powered submarines were "extremely noisy."[55] At a time when the U.S. Navy was producing the excellent *Los Angeles* and *Seawolf* classes, and when American submariners had benefited from years of Cold War undersea experience, the Chinese fleet was unable to send its ballistic missile submarine on a single operational patrol, and the rest of its fleet rarely if ever left shallow coastal waters. The RAND Corporation estimated that just 3 percent of Chinese attack submarines could be considered modern as of 1996.[56]

Yet after the 1996 Taiwan Strait crisis, when President Clinton dispatched two carrier strike groups in a show of force, the Chinese submarine fleet has rapidly improved to match the changes in its surface and aerial forces. In roughly the last twenty years, the PLAN has acquired more than thirty-five submarines of five classes, retired large numbers of obsolete submarines, and has "carved out of solid rock" a new base for its most modern submarines.[57] As of 2020 the Chinese fleet included six nuclear attack submarines, four ballistic missile submarines, and fifty diesel attack submarines, making it the largest undersea fleet in the Pacific.[58] RAND estimated that at least 66 percent of Chinese submarines can be considered modern, up from the previously mentioned 3 percent. These improved platforms are also venturing farther and farther out from Chinese coastal waters, and their longer patrols have included multiple deployments to the Indian Ocean, demonstrating the undersea fleet's improving readiness, training, and logistical support.[59] These rapid advances led longtime ASW expert James Bussert and Naval War College professor Bruce Elleman to posit, "PLAN submarines have transformed from the old Soviet imports to quiet, high-technology, and respectable indigenously designed diesel submarines (SSs) and modern nuclear-powered submarines (SSNs). A dramatic increase in Chinese diesel boat technology negates the comfortable stereotype of noisy old Soviet-style submarine copies clanking around China's coast."[60]

Despite these improvements, the Chinese submarine fleet still has numerous shortcomings and is not on par with the U.S. fleet. Its new *Shang*-class SSNs are "still relatively loud," and its older SSNs are "probably the noisiest nuclear submarines still in commission anywhere in the world," according to the Naval War College's China Maritime Studies Institute.[61] With limited ASW helicopters and MPA, the PLAN seems for the moment to be largely ignoring ASW operations and instead focusing its submarine fleet on "regional missions that concentrate on [anti-surface warfare] near major [sea lines of communication]," as the Office of Naval Intelligence (ONI) has concluded.[62]

The PLAN is also negatively affected by maintenance and logistical issues, which play a major role in submarine operations by keeping ships quiet and operational. PLAN Admiral Zhu Shijian has expressed envy of U.S. Navy development and management systems, and in 2012 he said that with respect to submarine logistics between the two nations, "The gap is very wide."[63] Despite these issues, PLAN submarines pose a serious threat to the U.S. surface fleet, one that will likely grow as Chinese technology, training, and procedures improve. Naval War College professor Lyle Goldstein and Naval Undersea Warfare Center analyst Shannon Knight wrote, "Submarines are still the sharpest arrow in China's quiver."[64] Four Naval War College professors, all experts on the Chinese military, opened their book, *China's Future Nuclear*

Submarine Force, by stressing the importance of submarines to the PLAN: "It is now widely recognized that submarines are the centerpiece of China's current naval strategy."[65]

Chinese submarines can be loosely sorted into three categories: obsolete submarines, capable nuclear submarines, and capable diesel submarines. Many of China's old submarine classes, such as the Romeo, Han, and Ming attack boats, are being retired and would struggle to survive against enemy ASW forces. The most capable of these classes is likely the Type 035 Ming class, but even these submarines are more than forty years old and cannot fire ASCMs, and in 2003 an apparent accident onboard one killed the entire crew.[66] Due to their limited capabilities and poor acoustic signatures, these submarines would be unlikely to pose a serious threat to American aircraft carriers in a conflict. However, as retired Rear Adm. Eric McVadon pointed out, the Chinese could deploy roughly twenty of these submarines in an effort to "complicate" the ASW picture, because they are "noisy but cannot be ignored."[67] The U.S. Navy could be forced to waste precious time and resources tracking and destroying these old boats, and they could be used to draw out U.S. submarines for attacks by other more capable PLAN submarines.

The small Chinese nuclear attack fleet is led by the Type 093 Shang class. ONI assesses the Shang to be comparable to the Soviet Victor III submarine class, which had improved systems, quieter noise levels, and ability to sail at high speeds.[68] Despite being much better than the old Han SSN class, the Shang is still reported to be vulnerable to long-range passive detection by numerous ASW platforms, and would likely be disadvantaged in deep water operations against the U.S. fleet.[69] The Shang class is capable of firing both active/passing homing torpedoes and wake-homing torpedoes as well as the advanced YJ-82 ASCM.[70]

After a decade-long pause in construction, China built four additional Shang submarines.[71] This improved Shang variant may have a flank sonar array, a towed array, and better quieting technology, which could make it a more formidable threat.[72] After this variant is complete, the PLAN is expected to begin construction on the Type 093B class of guided-missile submarines and the Type 095 SSN class, which ONI assesses "may provide a generational improvement in many areas such as quieting and weapon capacity."[73] These new designs, if they are able to combine the benefits of ASCMs and nuclear power with a reduced acoustic signature, may be better able to challenge U.S. surface fleets, but it is too early to tell.

However, the backbone of the Chinese submarine fleet is the diesel-powered attack submarines, including the Song, Yuan, and Kilo classes. All thirteen Song, seventeen Yuan, and eight of the twelve Kilo submarines are capable of firing advanced ASCMs.[74] ONI assesses them all to have acoustic signatures ranging from average to excellent.[75] The first Type 039 Song was launched in 1994.[76] It incorporates numerous foreign technologies including German diesels, French digital sonar, and also has an advanced seven-blade skewed propeller and anechoic tiles to absorb emitted sound. Professor Murray assesses the Song to be "the rough equivalent of a mid-1980s Western diesel submarine, which makes it a formidable, quiet submarine that will be very difficult to detect and locate, at least when the vessel operates on its batteries."[77] Songs can carry acoustic or wake-homing torpedoes as well as YJ-82 ASCMs that can be launched submerged, which was a first for the Chinese fleet at the time.[78]

In 2004 the PLAN launched the first Type 039A Yuan-class submarine, which has been described as a hybrid of the Chinese Song and Russian Kilo submarines. It is believed to be very quiet, have a seven-blade propeller as well modern periscopes and sonar, and can fire a variety of modern torpedoes and ASCMs.[79] ONI believes the Yuan to be China's most modern conventionally powered submarine.[80] What makes the Yuan so impressive, however, is its air-independent propulsion. AIP technology has been implemented on German, Swedish, and French submarines, and after years of rumors the Department of Defense confirmed in 2015 that the Yuan was also AIP capable.[81] Although AIP technology can take several forms, its greatest advantage is that the submarine only needs to snorkel roughly every two weeks instead of every one-to-three days, what is typical for conventional diesel submarines.[82] This reduced snorkel frequency is a massive advantage. When conventional submarines operate on the battery, they are exceptionally quiet due to their few moving parts, but their Achilles' heel is their need to expose a mast above the surface and operate noisy diesels for several hours. However, by considerably reducing the snorkeling frequency, AIP submarines like the Yuan greatly mitigate this weakness, and give ASW forces fewer clues spread out over longer periods of time. The Yuan's impressive weapons, improved quieting technology, and AIP system were likely part of the reason that Lyle Goldstein referred to the class as "one of the sharpest spears in China's maritime arsenal."[83]

The final Chinese submarine class is the Russian-built Type 877 and Type 636 Kilo. Two of the export variant Type 877 were delivered in 1995, but the other ten purchased are all of the significantly more capable Type 636 design. Eight of these ten came with the ability to fire the highly advanced 3M-54E Klub (SS-N-27B Sizzler) ASCM, a weapon with a reported 220-km range, terminal speed of Mach 2.9, and an ability to perform low-altitude evasive maneuvers to avoid Aegis defenses.[84] These weapons are especially dangerous because of the Type 636 Kilo's quiet operation, making it incredibly difficult for ASW forces to find it before it can launch. ASW expert James Bussert and Naval War College professor Bruce Elleman wrote, "Kilos are among the quietest and most modern ASW submarines in the world,"[85] and in 2009 ONI assessed the Kilo to be China's quietest submarine and on par with the new Russian St. Petersburg class.[86]

At first glance, the Chinese submarine fleet may appear to be unable to seriously challenge a well-protected aircraft carrier. Its few fast nuclear submarines may be too loud to survive long enough to penetrate ASW defenses, and its capable diesel submarines may be too slow to catch up and launch their torpedoes at a carrier. However, there is a good chance that the Chinese are overcoming this problem by focusing on anti-surface warfare by ASCM, the same strategy employed by much of its surface and aerial forces.[87] If the PLAN employed this strategy, its submarines could spread out and quietly wait for external cueing to launch a salvo of ASCMs.

This ASCM-centric strategy would mitigate many of the Chinese weaknesses while helping the PLAN avoid many of the U.S. Navy's strengths. By waiting for external cueing to launch their long-range weapons, Chinese submarines would not be forced to conduct high-speed, noisy evolutions in an attempt to find and close with fast carrier strike groups, thus minimizing the chances that U.S. ASW forces or submarines could detect and kill them. By relying on ASCMs instead of shorter-range torpedoes, Chinese submarines can also avoid the carrier's ASW defenses. It requires a great deal of training and skill to stealthily evade ASW MPA, helicopters, and surface escorts in order to launch torpedoes at carriers, but the Chinese could avoid this problem by relying on ASCMs launched in a safer environment far from the threat.[88] All of these factors would make it much more difficult for American ASW forces to find and destroy a quiet Song, Yuan, or Kilo that did not need to reveal itself until it launched its missiles, and this hunt would be made even more challenging by the acoustically difficult waters surrounding China. If PLAN submarines remained within the first island chain, then the shallow water and heavy shipping traffic would shorten passive and active sonar ranges, and vulnerable U.S. ASW aircraft like the P-8, P-3, and MH-60R may be unable to conduct submarine searches when threatened by Chinese fighters operating from bases on the mainland and islands of the South China Sea.[89]

While there would certainly be major challenges with this strategy, and there is limited proof that it is the course of action the Chinese would select, it does appear a likely and attractive option from the PLAN point of view. A Chinese fire control textbook, written by a PLAN captain who teaches at a submarine academy, says that the PLAN is working to "equip attack submarines with long-distance, high-speed (Mach), low-altitude flight, high-accuracy, strong interference-resistance [sic] antiship missiles with the combat capability to attack enemy surface ships from mid- to long-range." The captain also wrote, "Under modern combat conditions, the main combat method for attack submarines is to fire antiship missiles from underwater to attack enemy ships."[90] Captain Patton discussed the likelihood of this Chinese strategy, writing, "U.S. planners may be deluded into believing that China will surge large numbers of submarines against carrier and amphibious battle groups for torpedo attacks. In fact, Beijing would be wiser to plan to dilute the American ASW efforts with large numbers of boats deployed, but not attempting to penetrate ASW defenses. Submarine attacks against surface targets would instead by directed by third-party targeting fire with long-range antiship cruise missiles (ASCMs) from submarines in relatively safe locales awaiting targeting data."[91]

Facing off against all of these undersea threats are U.S. SSNs and the carrier strike group's formidable ASW defenses. As defense analyst Robert Haddick wrote in Fire on the Water, "The U.S. fleet's best trump card is its attack submarines."[92] These submarines would likely inflict heavy losses on PLAN surface ships and submarines in a future conflict, based on their excellent design, the training and experience of the officers and crews, and China's lack of "either a robust coastal or deep water anti-submarine warfare capability," according to the Department of Defense.[93] The various U.S. air and surface platforms could also be expected to give good protection in the short- to mid-range, as discussed earlier in the chapter. Finally, the ability to locate, identify and track a carrier strike group and then pass that information to waiting submarines would certainly be challenging and it is unclear if China has that capability.[94]

Despite these defenses and U.S. advantages, there is still a great deal of evidence indicating that large aircraft carriers could be vulnerable to Chinese submarines. Most of the problem stems from the basic physics of underwater acoustics. As Chinese submarines get quieter, it will only become increasingly more difficult to detect them, never mind track and destroy them. As these platforms' broadband and narrowband signatures get smaller, the ranges at which they can be effectively detected will continue to decrease. As that range decreases, it will either require more time to search the same area, or more friendly platforms to accomplish that search in the same amount of time. At the same time, the area that ASW forces must search is increasing due to submarines' improved weapons ranges. In simplest terms, ASW forces are being forced to search for a needle that is getting smaller in a haystack that is getting larger.

To confound the problem, there are a diminishing number of searchers and they often have other tasks to accomplish. As of 2020, there are twenty-seven U.S. fast-attack and guided-missile submarines in the Pacific Fleet.[95] Even if two-thirds of them could be on the frontlines in a conflict, that leaves just eighteen submarines to conduct a wide variety of high-priority tasks like ASUW, ASW, strikes against land targets, intelligence collection, and special operations missions. Devoting a significant portion of those theoretical eighteen submarines to protect carrier strike groups would come at a massive opportunity cost.[96] Similarly, U.S. surface escorts employing their powerful SQQ-89 ASW combat system and operating with MH-60R helicopters very well may be able to effectively defend against enemy submarines, but as MIT's Dr. Cote points out, the problem is that these escorts "are also primary fleet air and missile defense assets, as well as cruise missile shooters, and they can only be in one place at one time."[97] Finally, the replacement of specialized ASW air platforms like the S-3B Viking and SH-3 Sea King with multi-purpose aircraft like the MH-60R have negatively affected the carrier's ASW defenses. These substitutions led Norman Polmar to write in his definitive history of the aircraft carrier, "The carrier-based ASW capabilities have been severely reduced."[98]

Many of these problems are mitigated in the Flex Fleet. As submarines get more difficult to detect, more searching platforms—friendly aircraft, surface ships, and submarines—will be needed to find and destroy them in a reasonable amount of time. By adding eighty-one *Constellation*-class frigates with helicopters and 110 Steel-class corvettes, the Flex Fleet achieves a 58 percent increase in ASW-capable surface ships over what the U.S. Navy is planning to build. Those additional frigates and corvettes would perform numerous ASW missions, freeing up the more valuable *Arleigh Burke*–class destroyers and *Virginia*-class submarines for higher-priority missions. The Flex Fleet is built to maximize the use of the undersea realm instead of treating it as another carrier vulnerability.

In 2015 the think tank RAND Corporation developed a model to study Chinese submarines' ability to attack U.S. aircraft carriers, as part of a comprehensive comparison of both nations' militaries. In this model, RAND used estimates of acoustic signatures, available forces, weapons ranges, and numerous other factors to arrive at estimates of the number of potential engagements Chinese submarines might achieve against U.S. aircraft carriers. In order to study the trends over time, it repeated this analysis using data from 1996, 2003, 2010, and predictions for 2017. Following this in-depth study, RAND analysts came to the following tentative conclusions:

> First, the Chinese submarine fleet has made major gains relative to U.S. defensive capabilities. Under any single set of assumptions assessed within the model, the number of expected potential engagements by Chinese submarines against U.S. carriers increases by more than an order of magnitude (and, in some cases, by more than 20 times) between 1996 and 2017.

Second, cueing could also substantially improve the Chinese ability to engage U.S. targets. Daily cueing increases the average number of engagement opportunities by a factor of five to eight, depending on other assumptions. . . .

Third, the modeling results suggest that not only is the threat increasing rapidly, but it has also become significant in absolute terms, a fact that may have implications for how the United States employs its carriers. Even without cueing, Chinese submarines might have close to an even chance of engaging a single U.S. carrier over a seven-day period. . . . Given the cost, number of personnel, and symbolic importance of U.S. aircraft carriers, this level of risk could prompt U.S. commanders to hold carriers back until areas closer to China could be sanitized by U.S. anti-submarine assets.[99]

Chinese submarines appear to pose another serious threat to the aircraft carrier. Taking advantage of their quiet designs, advanced weaponry, the difficult acoustic environment of the Chinese near seas, and improving external cueing, the Chinese submarine fleet will likely be a "perplexing, long-term issue" for the U.S. Navy.[100]

Unfortunately, the Chinese submarine fleet is not even the most capable potential foe the U.S. Navy may have to confront.[101] The Russian undersea fleet has a great deal more experience, better technical expertise and infrastructure, and submarines that are more advanced and armed with deadlier weapons. The Kilo and Sizzler ASCMs—arguably the best platform and weapon in the Chinese arsenal—are Russian exports, evidence of how far the PLAN still has to go. Most importantly, while Chinese nuclear submarines are still somewhat loud, Russia has produced numerous nuclear fast attacks that are much quieter and more capable. For example, in 2009 ONI evaluated the Chinese Shang submarine and found it to be louder than seven entire classes and variants of Russian nuclear submarines, including the Victor III, Akula I and II, Oscar II, and the new *Severodvinsk* class.[102] These nuclear attack boats, as well as the new St. Petersburg class of diesel-electric submarines that improve on the already formidable Kilo, are likely even more lethal than the Chinese submarine fleet.[103] Russia has compounded the problem through its willingness to sell submarines to other navies. For example, in addition to

Russia and China, Poland, Romania, India, Algeria, Iran, and Vietnam all operate Kilo submarines.[104]

In 2021, the U.S. Navy's undersea surveillance commander, responsible for what was known as Sound Surveillance System (SOSUS) arrays and is now the Integrated Undersea Surveillance System, summed up the Russian challenge by saying, "The submarines that they are putting out into the ocean are increasingly capable and increasingly quiet and becoming harder and harder contacts to find."[105]

MINE WARFARE TODAY

In addition to submarines, mines pose a significant and often overlooked threat to modern warships. When considering defense budgets and infrastructures, mines are extremely easy to design, develop, and deploy, making them an attractive option for overmatched navies looking to defend against the power of the U.S. Navy. The U.S. Navy's "2009 21st Century Mine Warfare" report identified the allure of mines, stating, "These 'weapons that wait' are the quintessential global asymmetric threat, pitting our adversaries' strengths against what they perceive as our naval and maritime weaknesses."[106] Complicating the historic mine problem are new technologies enabling the development of "smart mines" which are even more difficult to defend against and which pose an even greater threat to all ships. As Vice Adm. Michael Connor, a former Submarine Force commander, wrote about those technological advances, "The torpedo of the future and the offensive mine of the future will be hard to distinguish."[107]

Today an estimated sixty navies have roughly 1 million mines of 300 different types, not including the U.S. Navy's inventory.[108] One of the world's leaders in mine technology and production is the People's Republic of China (PRC), which has an estimated inventory of 50,000 to 100,000 mines.[109] In a description very similar to those discussing the PRC's air, surface, and submarine fleet improvements, ONI stated, "During the past few years, China has gone from an obsolete mine inventory, consisting primarily of pre-WWII vintage moored contact and basic bottom influence mines, to a vast mine inventory consisting of a large variety of mine types such as moored, bottom, drifting, rocket-propelled, and intelligent mines."[110]

The U.S. Naval War College's China Maritime Studies Institute performed an in-depth analysis of China's mine inventory and concluded that it "is not only extensive but likely contains some of the world's most lethal [mine warfare] systems. Indeed, China is on the cutting edge of mine warfare technology and concept development, and it already fields systems that advanced nations—the United States, for one—do not have in their arsenals." The analysis continued, "Beijing's military modernization program is a comprehensive effort, striking in both breadth and focus. Chinese [mine warfare] is noteworthy because it is one of a few warfare areas that could, in conjunctions with other capabilities, suddenly and completely upset the balance of power in the western Pacific."[111] Because of the threat mines pose to U.S. aircraft carriers, this section briefly reviews the Chinese mine warfare capability as a proxy for any potential U.S. adversary.

Sea mines are typically classified by their position in the water, their method of delivery, and their method of actuation.[112] Considering final position in the water, mines can be moored, bottom, or floating. Moored mines are positively buoyant, which limits their explosive charge, and held in place by an anchor or cable. They can be utilized in very deep waters, but their moorings can be cut by mechanical sweep gear. Despite their relative simplicity, these early mines can still be highly effective; both USS *Tripoli* and USS *Samuel B. Roberts* were damaged by moored mines. A significant portion of the PRC's mines are moored variants, and although older, mines like the acoustically triggered M4, with a 600-kg warhead, could inflict very serious damage on even large targets like aircraft carriers.[113] A more dangerous subset of moored mines are known as rising mines, which have a torpedo or rocket-propelled explosive that is released from its mooring and accelerates up toward its target before detonating. These mines can be laid in significantly deeper water. For example, China has purchased the Russian-built PMK-2, which is acoustically triggered and armed with a 110-kg warhead and is reportedly capable of being deployed in waters up to two thousand meters deep.[114]

Bottom mines rest on the seabed, or are buried slightly under it, allowing them to be negatively buoyant and thus have larger explosive charges. In order to have the required sensitivity and be close enough to passing ships to cause damage, bottom mines can typically only be used in waters less than two hundred feet deep.[115] Bottom mines are much more difficult to detect than moored mines because they blend in with the seabed and are also more difficult to sweep because there are no chains that can be cut to cause the weapon to float to the surface. Chinese bottom mines include weapons like the C-1 1000 and EM-12, both with seven-hundred-kilogram warheads.[116] Although unusable in deep water, they could be attractive options for the shallow waters of the Chinese near seas. Submarine-launched mobile mines, such as the Chinese EM-56, are another dangerous type of bottom mine. The weapon propels itself to areas often inaccessible to the launching submarine, such as enemy ports or very shallow chokepoints. Once it arrives at its designated location, it shuts down its propulsion system and sinks to the bottom. The EM-56 has a 13-km range; an acoustic, seismic, and pressure triggering system; and a 380-kg warhead.[117]

The final category of mine positioning is floating mines. If they come free of their moorings, they become known as drifting mines and are dangerous weapons that go wherever the ocean carries them. Drifting mines are generally illegal but are still used today.[118] The Chinese have allegedly manufactured at least three variants of drifting mines, including the small Piao-2, which is designed to attack small and medium surface ships and can be easily deployed by small fishing boats.[119]

Mines are also classified by their deployment platform, which can include aircraft, warships, minelayers, submarines, merchants, and trawlers. The variety of platforms and relative ease of laying mines make it difficult to prevent their use, especially considering stealthy options like submarines and civilian shipping. This stealth means that the first evidence of the presence of mines could be their detonation, such as when a Libyan ferry discretely laid mines in the Red Sea and Gulf of Suez for two weeks in 1984, resulting in underwater damage to twenty-three vessels.[120]

The vast majority of the Chinese surface fleet can carry mines, including destroyers of the Sovremmeny, Luhai, Luhu, and Luda classes, Jianghu-class frigates, and various patrol craft and minesweepers.[121] An estimated 150 aircraft, including the H-6 bomber, can also carry mines, which gives the PRC the ability to rapidly lay minefields or replenish older ones.[122] In addition, many experts believe the Chinese submarine fleet is heavily invested in mine warfare. All Chinese submarine classes can carry mines, which may provide an explanation for the retention of obsolete classes, such as the Romeo.[123]

The final major mine delivery platform, civilian shipping, is likely the most difficult to defend against. The best mine defense is to ensure they are never laid, but this task would be essentially impossible against thousands of Chinese trawlers and merchants that sail the Chinese near seas every day. These civilian ships, part of the PAFMM, could discretely lay mines at night and often with equipment already installed on board.[124] The PRC is reportedly making extensive preparations to use the PAFMM for civilian ship minelaying, including numerous robust training exercises.[125] In addition, as maritime expert and *Naval War College Review* editorial board member Dr. Scott Truver pointed out, the massive state-owned China Ocean Shipping Company, known as the COSCO Group, could also be used to lay mines around China and in foreign ports.[126] Even if many civilian ships could not secretly lay large numbers of mines, they could still be used to replenish depleted minefields or deploy these weapons in select chokepoints and harbors.

Finally, mines are also classified by their triggering mechanism. They can be detonated by contact, on command, or on influence signatures such as magnetic, acoustic, pressure, electric field, or seismic.[127] Modern mines can combine these triggers or add other features making their detection and neutralization more difficult. Some classes have time delay or ship counting features, which would allow them to be laid before hostilities were initiated. Minesweepers often attempt to safely trigger a mine's detonation by creating false signals, but many mines now have multiple triggers which all must be satisfied for the weapon to explode. Furthermore, some mines can be remotely controlled via acoustic signals, meaning they could be temporarily disabled while minesweepers were in the area, only to be turned back on when targets appear.[128] All of these capabilities complicate mine countermeasure efforts and force those assets to individually hunt and disable mines instead of sweeping large numbers of them at the same time. In addition, the PRC is reportedly adding new technology to its old weapons, converting them into "smart" or "intelligized" mines that can be programmed to only attack certain ship types and that are "virtually impossible to sweep," according to the Naval War College's China Maritime Studies Institute.[129]

To defeat the mine threat, mine countermeasure (MCM) operations are broken into offensive and defensive efforts. Offensive MCM seeks to break one of the links in the enemy's minelaying kill chain by destroying mines in production, storage, transportation, or laying.[130] This route is the best option for MCM forces, because it removes the need to find, classify, and neutralize mines once they are deployed. It also gives MCM forces the opportunity to destroy hundreds of mines at once, such as in a warehouse or on a transporting ship, rather than having to destroy each mine individually. With its vast intelligence and surveillance networks and powerful strike options, the U.S. Navy would likely prove capable of conducting some successful offensive MCM operations.

Despite this potential, there are several issues which make offensive MCM incredibly difficult to consistently excel at, and which would weaken the Navy's ability to protect the aircraft carrier and the rest of its fleet. Tracking the construction and movements of large surface warships, submarines, and aircraft is challenging but relatively straightforward, but gaining intelligence on the location, inventory, and capabilities of small mines is significantly more difficult. Unlike many other larger and more expensive weapons, mines can be designed, constructed, and stored in nondescript locations far from conventional military sites.

Even if the Navy can gather that intelligence, any offensive MCM operation is "frequently a short notice, time critical event," as stated in the Navy's publication on mine warfare. The manual continues, "The determination that loading is in progress must be followed within a matter of hours by the complete sequence of strike planning, approval, and execution if the offensive MCM operation is to be successful. Delay may result in striking after the mine movement is complete. Complicating the problem is the likelihood that if hostile intent exists, the transfer of mines will be carried out surreptitiously in darkness or using deceptive methods."[131] Additionally, it may be politically impossible to conduct preemptive attacks on mine forces. Attacks may not be allowed on a country's mainland, such as against Argentina during the Falklands Conflict. Additionally, mines may be deployed before hostilities begin.

The final major challenge of offensive MCM deals with quantity. The U.S. Navy, with dozens of highly capable strike platforms, could relatively easily find, track, and destroy a minelayer before it accomplishes its mission. The challenge is not in breaking that one minelayer's kill chain, but in breaking the kill chain of hundreds of minelaying platforms in a short period of time. That task becomes even more challenging due to the diversification of the entire minelaying network. Because of their small size and simplicity of use, mines can be built, transported, and stored in a wide variety of locations ensuring there is no single link that is vital to all the kill chains. By using warships, submarines, and aircraft, there are numerous ways the mines can be laid such that each must be individually found, tracked, and destroyed before U.S. ships can complete their mission. Adding civilian shipping to that minelaying fleet, which can blend in with hundreds of other innocent ships, complicates the challenge even more.

Historically, the U.S. Navy has failed in offensive MCM operations against opponents such as the North Koreans, Libyans, and Iraqis, none of which had a credible navy. The U.S. Navy's ability to stop minelaying operations today, against similar foes or more formidable adversaries such as China, Russia, or Iran, is likely to be similarly poor.

If the Navy cannot prevent the enemy from laying mines, it must shift to defensive MCM, which includes operations to search for, plot, and either avoid or neutralize deployed mines. The first step in defensive MCM is to find the mines, an exceedingly difficult task simply due to physics. This chapter reviewed how difficult it is to detect submarines using scouting methods including active and passive sonar, radar, magnetic anomaly detection, electronic support measures, and visual searches. Unfortunately, finding mines is significantly harder. Submarines are quiet, but mines are silent, so passive sonar is not an option. Mines transmit no energy and are below the surface, so electronic support measures, radar, and visual searches are all ineffective. Resolving the magnetic distortion from a large submarine hull is difficult, but mines are a fraction of the size and cannot be effectively detected using magnetic means.[132]

The primary, and realistically only, way to detect mines is active sonar. Because mines are so much smaller than submarine, the sonars must employ a higher frequency to achieve the required resolution. However, these high frequencies suffer a great deal of attenuation and scattering in the ocean, limiting their range, and thus requiring a great deal of time for MCM forces to search large areas.

If the Navy's MCM forces fail to detect mines either through intelligence or search, the consequences can be severe. As the Navy's Mine Warfare doctrine states, if that initial mine detection fails, "Initial reconnaissance will most likely be performed by unprepared merchant or naval ships and the mining incidence will be documented by damage reports."[133]

Once the mines are found, they can be avoided or neutralized, depending on their location. If simple avoidance is not an option, the Navy must either sweep or hunt the mines. Minesweeping is the process of clearing large numbers of mines using mechanical or influence systems that reveal or trigger any mines in the area. Mine hunting is the process of finding the exact locations of individual mines and then removing, neutralizing, or destroying each of them.[134]

To complete these active MCM operations, the Navy relies on its MCM triad: surface MCM ships, MCM helicopters, and explosive ordnance disposal (EOD) teams. That triad is faced with a difficult and time-consuming task. In order to sweep mines, MCM forces must mechanically cut mines' chains, causing them to float to the surface where they can be destroyed. However, bottom mines have no chains to be cut, forcing MCM forces to replicate the mines' triggering criteria. This option is becoming more difficult as mines are built with simple logic to ensure they only trigger for actual targets.[135]

If mines cannot be swept, they must be individually hunted. This method is preferred because of its thoroughness, but it requires a great deal of time and resources. Three professors from the Naval War College's China Maritime Studies Institute discussed the challenge with mine hunting, writing that it is a "time-consuming and arduous process, requiring not only extremely accurate bathymetric mapping but also the painstaking investigation of every minelike object on the seabed in the area of concern. This requires advanced, expensive technology, specialized training, and high levels of localization accuracy."[136]

Mines' inherent advantages of physics and numbers mean that MCM forces looking to protect the aircraft carrier and any other warship must invest incredible amounts of resources and time to neutralize them. Even after minefields are located, it takes several hours for each mine to be localized, identified, and neutralized before moving onto the next one.[137] The entire process becomes much more difficult and time-consuming if MCM forces are under threat of attack.

The difficulty associated with MCM was clearly demonstrated multiple times in the Persian Gulf. When mines were discovered during the Iran-Iraq War, six countries sent a total of twenty-five MCM ships, six MCM helicopters, and numerous EOD teams to clear the shipping lanes. They worked for almost two years to clear just twenty-six mines, yet despite this effort nine ships were hit by mine explosions, two of which sank.[138] Just a few years later during Operation Desert Storm, two U.S. warships hit mines in an area declared to be mine-free.[139] Former Chief of Naval Operations Adm. Frank Kelso summed up the difficulty associated with mine warfare when he responded to these events: "I believe there are some fundamentals about mine warfare that we should not forget. Once mines are laid, they are quite difficult to get rid of. That is not likely to change. It is probably going to get worse, because mines are going to become more sophisticated."[140]

In a *Naval War College Review* article, Dr. Truver discussed the difficulty associated with MCM, and why the U.S. Navy often chooses to focus on other, more glamorous missions. He wrote,

> In a way, Big Navy's indifference, if not hostility, to investment in MCM is not without merit. Looking objectively at mine-hunting technology versus advanced mine technology, the Navy cannot have any real confidence that a quick and effective in-stride mine-clearing capability in a non-benign environment will ever be achieved. Post-Desert Storm, the world's best MCM capabilities were for the most part pitted against relatively ancient mines. The clearance rate was painstakingly slow and could only be achieved in a totally benign environment. Following the end of Desert Storm hostilities, an international MCM force needed some two years to declare Persian Gulf sea-lanes and ten mine-danger areas to be mine free.[141]

Additionally, the Naval War College's China Maritime Studies Institute concluded that "at present, the prospects for American MCM forces rapidly countering Chinese [mine warfare] are not promising."[142]

Mines pose a serious threat for the U.S. Navy's aircraft carriers, as well as all its other warships. It takes a great deal of intelligence, presence, time, and energy to prevent the laying of mines, which is relatively easy for the enemy and can be done by aircraft, warships, minelayers, submarines, merchants, or trawlers. Once they are laid, it takes a great deal of time and energy to find, plot, evade, or neutralize these weapons. They can be placed in difficult acoustical environments, buried in the seabed, employ microprocessor control to resist minesweeping efforts, or be constructed with nonmetallic casings to reduce their sonar and magnetic signatures. If insufficient time and energy is invested in mine countermeasure efforts, the alternative is to risk damaging or losing a $12 billion asset and the success of the current mission, all because of a $20,000 weapon. That weapon is a threat to the rest of the ships of the U.S. Navy, as well, but centralizing so much of the fleet's firepower and capabilities on a single ship makes it too big to fail and amplifies the impact a single mine can have. The carrier does not even need to be damaged; the mere threat of a minefield may be enough to deter commanders from deploying the carrier to where it is needed.

TOO BIG TO SINK

This chapter has analyzed the threat submarines and mines pose to today's aircraft carrier. That analysis has been clouded by the fact that the "undersea arena is the most opaque of all warfighting domains," as Admiral Connor put it.[143] However, there is strong evidence that the aircraft carrier's two most basic ASW options—fight or flight—are less attractive today than ever before, and only becoming worse. The range at which submarines can be detected is shortening while the submarines' weapon ranges are increasing, a combination of factors that strain the carrier's defenses like never before. That problem is further complicated by the presence, or mere threat, of naval mines.

The single most important factor in undersea warfare is sonar, and as enemy submarines get quieter both their warfighting and deterrence capabilities grow. Quiet, effectively employed submarines are extremely difficult to detect and track, which gives them more time and better opportunities to launch attacks. Admiral Connor discussed the undersea challenge, saying, "It is easier to track a small object in space than it is to track a large submarine, with tremendous fire power under the water."[144] Even without launching weapons, these undersea threats can act as a "paranoia multiplier" and may force political and military leaders to rethink whether sending a ship with a crew of five thousand Americans into harm's way is worth the risk.[145] Captain Hendrix summarized the situation, writing, "Perhaps no place poses a greater threat to the current U.S. force structure or suggests the greatest potential for improvement in a future Navy than the underwater environment and the vessels that populate it."[146]

The Flex Fleet would be one force structure that could reduce paranoia and better control the undersea. Its *Constellation* frigates and Steel corvettes would supplement *Arleigh Burke* destroyers and *Virginia* submarines to find and destroy enemy submarines faster than is possible with today's fleet structure. The Flex Fleet would further reduce paranoia by cutting the aircraft carriers' responsibilities, thus allowing them to operate farther out to sea, where enemy submarines and mines would be less likely to affect them. Most importantly, more friendly submarines would be available for offensive operations, where they are truly in their element. As Capt. Robert Rubel wrote, certain sectors of the seas, "are becoming a no-man's-land for surface ships. Whether or not submarines ought to be considered capital ships is beside the point; the carrier will likely not be one."[147]

The submarine and mine forces of nations like China, Russia, and Iran are formidable, but the U.S. Navy still has a sizeable advantage because of its superior technology, training, and experience. Unfortunately, that advantage—like all competitive advantages—is likely to be eroded over time. The U.S. Navy should seize the opportunity to develop a powerful and dispersed fleet that can still accomplish its mission despite the presence of enemy submarines and mines. The other option is to continue to rely on a platform that is more vulnerable to undersea threats than ever before.

Reviewing all the threats facing the aircraft carrier, many are so dangerous because they have physics and sheer numbers on their side. Without considering exact weapons, systems, or specifications, physics makes it easier to hit a carrier with an ASCM than it is to hit an ASCM with an interceptor missile. Physics makes it easier to hit a carrier moving in two dimensions at thirty knots with an ASBM than it is to hit an ASBM moving outside the atmosphere in three dimensions at ten times the speed of sound. The physics of underwater acoustics makes it easier for a deep quiet submarine to find an aircraft carrier and its escorts than it is for those forces to find that stealthy submarine. Similarly, physics makes the detection, identification, and destruction of small, silent objects moored in place or resting on the seabed incredibly difficult, regardless of the exact mine countermeasure systems in operation today. Any navy attempting to avoid or destroy each and every one of these threats is starting off at an advantage simply due to physics.

To complicate the problem, many of these weapons have a quantity advantage. A strike group has only one carrier, but hundreds of ASCMs and dozens of ASBMs can be fired at it, and even if only a couple of each hit their target, the enemy will have scored a massive success. Dozens of submarines armed with dozens of torpedoes and missiles can prowl the seas, yet it only takes one to make a successful attack. Mines take this numerical advantage to a new level. Enemies can somewhat easily lay hundreds or thousands of mines in a relatively short period of time, secure in the knowledge that if just one or two hit the carrier they will have accomplished their mission. Meanwhile, the carrier must avoid or destroy essentially every threat to guarantee its ability to complete its mission.

These weapons threaten all surface ships and all navies, not just U.S. aircraft carriers. The difference is the aircraft carrier is so important to the U.S. Navy and the nation that its loss would be catastrophic, leading to unintended consequences in how the ship must be operated and defended.

The aircraft carrier remains by far the most important ship type in the U.S. fleet. Nearly all major mission areas are contingent on carriers, from strike warfare to fleet air defense; as Cdr. Phillip Pournelle, an alumni of the Department of Defense's renowned Office of Net Assessment, wrote, "All fleet operations are variations on the theme of carriers delivering ordnance," and that as a result, "the main, and realistically only, striking arm of the U.S. Navy is the aircraft carrier. By design, our navy is overly dependent on this one central platform for its combat effectiveness."[148] The carriers' warfighting importance is further amplified by their small numbers, with the U.S. Navy typically only having ten to eleven in commission at a time, with several unavailable due to extended maintenance periods. That importance is further increased by the significant financial, human, and political value of each carrier. Encapsulating more than $12 billion, carrying five thousand people and serving as a symbol of American might for the entire world, the aircraft carrier has an almost priceless value, and likely represents more of any fleet's firepower, people, and prestige than any capital ship in history.

Because of that outsized importance, the carrier's loss may be disastrous. During the 2008 financial crisis the theory emerged that certain financial institutions were "too big to fail." Some experts considered these firms to be so large and entwined with other companies that their closure would be catastrophic to the entire economy. In today's Navy, the aircraft carrier has become "too big to sink." When it functions as designed, it is an extremely powerful platform that has remarkable flexibility and economies of scale. But carriers are crucial to so many of the fleet's missions that if the enemy can defeat them, the results may be catastrophic for both the Navy and the nation. The loss of 10 percent of the Navy's destroyers or submarines would be painful and have serious effects, but would not fundamentally alter the fleet's warfighting ability. The loss of 10 percent of the carrier force—a single ship of $12 billion with five thousand Americans serving as the most recognizable symbol of U.S. military superiority—would send shockwaves around the world.[149]

Because the aircraft carrier cannot be lost, there are massive unintended consequences regarding its use and defense. In the 2008 financial crisis, the U.S. government took unprecedented steps to save companies that were "too big to fail," believing that drastic actions were justified to avoid the severe consequences if those businesses collapsed. Today, the U.S. Navy is forced to take extraordinary measures to provide a perfect defense for the aircraft carrier. As Captain Hendrix wrote,

> The rise of these threats over the past three decades has forced the Navy to emphasize the defensive capabilities of the carrier force, giving rise to the "anti" warfare commanders (antisubmarine, antisurface, and antiair). This emphasis on defensive capabilities occurred even as the effectiveness of the carrier's striking power has noticeably waned.
>
> The decisions that led to the current strategic condition have left us with a force that must operate at increased range from our adversaries in order to be safe (and preserve our expensive platforms), even as our

striking arm has decreased in its combat radius over time. Hence we find ourselves in a circular argument reminiscent of the late Adm. Hyman Rickover, that "I must defend my force, Sir, so that I can defend my force." The [carrier strike group] is, remarkably, a construct that can operate effectively only in a permissive environment, or be committed to an anti-access environment only under the most extreme conditions when national interests compel leadership to risk what amounts to a significant percentage of the Navy's annual budget in a single engagement.[150]

The mission has become the protection of the carrier rather than attacking effectively first. Yet the perfect defense required for a nearly priceless ship has rarely if ever been possible throughout naval history. Perhaps it was feasible in the early 1990s, with no credible peer threats at sea. Today, in the Age of the Missile, in the time of advanced submarines, and with the dual threats of China and Russia, the chances of effectively defending the carrier are smaller, and the chances of providing the necessary perfect defense have plummeted. With so much riding on the carrier, even a very good defense is likely not good enough.

Acknowledging this situation, the Navy's first option is to maintain its force architecture and invest more resources into carrier defense. It can continue to pour money and effort into defensive systems that are always going to be disadvantaged by physics and sheer numbers. More Aegis escorts, more defensive fighter squadrons, more interceptor missiles, and more submarines providing defensive services would all provide an incrementally improved defense for the carrier and improve its chances of success in battle.

Yet to provide a fundamental improvement in the fleet's warfighting abilities, the Navy's second option is to restructure the fleet by shifting missions away from the aircraft carrier. By moving to a structure like the Flex Fleet, the carrier can focus on missions it does best and be freed from other missions it is ill-suited for, enabling the use of smaller, less expensive, more numerous platforms that are not "too big to sink." Simultaneously, the rest of the fleet could reduce its defensive burdens, shift more focus to attacking effectively first, and form a stronger overall offensive network.

Two admirals, Robert Natter and Samuel Locklear, advocating for the enduring importance of the carrier, wrote, "Although often criticized for its expense and questioned for its relevance in the 21st Century, we still haven't found anything better than an aircraft carrier to uniquely conduct the wide spectrum of missions ranging from humanitarian relief to full scale conflict."[151] They are right in that there is no single ship more capable than the aircraft carrier, but the more useful standard is whether there are better fleet structures than one wholly reliant on them. In a structure like the Flex Fleet, each ship is individually weaker and less survivable than a *Ford*-class carrier, but the overall fleet is stronger.

PART III

SEIZING OPPORTUNITIES

Observe also that changes of tactics have not only taken place after changes in weapons, which necessarily is the case, but that the interval between such changes has been unduly long. This doubtless arises from the fact that an improvement of weapons is due to the energy of one or two men, while changes in tactics have to overcome the inertia of a conservative class; but it is a great evil. It can be remedied only by a candid recognition of each change, by careful study of the powers and limitations of the new ship or weapon, and by a consequent adaptation of the method of using it to the qualities it possesses, which will constitute its tactics. History shows that it is vain to hope that military men generally will be at the pains to do this, but that the one who does will go into battle with a great advantage.[1]

—Rear Adm. Alfred Thayer Mahan, USN (Ret.)

CHAPTER 8

100,000 TONS OF INERTIA

It may be stated in general terms that most arguments in favor of fundamentally new weapons have failed except those that resulted in shedding the blood of the unbelievers; that defeat alone has been accepted as a final demonstration.[1]

—Adm. William Sims, USN

The aircraft carrier is history's most powerful warship and the ultimate symbol of U.S. power. It has led the nation into battle around the world, from Midway to the Middle East. Despite that impressive history, there will come a day when the nation decides to stop relying on it. Just as the Navy must determine if it should stop building aircraft carriers, it must also determine if it could stop building them when it wants to.

There are numerous organizations and individuals with strong incentives to advocate for carriers—they won't let go easily. Fortunately, the Navy has a long history of applying the right forces to improve itself in the face of institutional and individual resistance. It will take a massive force to overcome the inertia of the 100,000-ton nuclear-powered aircraft carrier.[2]

INSTITUTIONAL INERTIA

Upton Sinclair wrote, "It is difficult to get a man to understand something, when his salary depends upon his not understanding it."[3] Today, a lot of salaries depend on the aircraft carrier, meaning—whenever the day arrives—it will be difficult to convince numerous key constituencies that the time has come to shift away from a carrier-centric fleet. These groups are varied but fit into three broad categories: the military, the defense industry, and the government.

The Military

Within the Navy, the aviation community obviously has the most to lose from the end of carriers. It would lose its primary operating platform, even as aviators already face the "twilight of manned flight."[4] Reducing reliance on carriers would shift many missions to surface ships and submarines, transforming the Navy's budget and upsetting the power balance among warfare communities within the Department of the Navy. Additionally, halting production of nuclear-powered aircraft carriers would cut the number of carrier and air wing command billets, restricting the path for aviators to reach flag rank.

But the aviation community would not be the only group to lose out. The entire community of nuclear-trained surface warfare officers would eventually cease to exist. In addition, the Naval Nuclear Propulsion Program—also known as Naval Reactors—would lose approximately 20 percent of its nuclear power plants. That portfolio reduction would restrict its influence to the submarine force alone, against its current role overseeing more than 40 percent of the Navy's major combatants.[5] As a result, it would only have influence over the selection and advancement of submarine personnel, losing its ability to affect the surface and aviation communities.

The Navy itself has strong incentives to keep building aircraft carriers to prevent a possible shift of strike missions, along with a corresponding reallocation of budget resources, to the Air Force.

The Defense Industry

Defense contractors also have strong motives to continue to build aircraft carriers. The defense industry often seeks to maintain large, stable contracts in a justifiable effort to avoid the costs associated with transitioning between complex projects. That stability is crucial, as it is not financially feasible to employ a large, skilled workforce with just intermittent work.[6]

To protect itself, the military-industrial complex musters an impressive lobbying effort. Entire organizations, such as the Aircraft Carrier Industrial Base Coalition (representing more than two thousand companies), advocate for the "importance of our nation's aircraft carriers" and emphasize the importance of a "stable industrial base."[7]

Retired flag officers working for defense contractors bolster those efforts, and they are present in large numbers. For example, as of 2022 all three of the most recently retired CNOs work for large defense contractors.[8] This enables contractors to put up a stiff fight when active-duty leaders attempt to change production priorities. In one case, 128 senior retired officers wrote a 2019 letter to Congress opposing a Department of Defense budget request that included just six fewer F-35s than originally planned. Lockheed Martin organized the letter, and 50 of the senior officer signatories may have had conflicts of interest.[9] That same year, the Army attempted to reduce purchases of an advanced variant of Chinook helicopters to instead buy two new vertical-lift aircraft. Even though then–secretary of the Army Mark Esper said, "Boeing . . . expressed support for our modernization strategy, and said they would support our budget," Boeing instead fought the cuts and succeeded in having them reversed.[10] If the services cannot cut small numbers of F-35s and Chinooks without such a response, imagine the uproar that would likely result over a plan to cancel the *Ford* class of carriers, a contract an order of magnitude larger.

Part of that uproar would likely come from Huntington Ingalls Industries (HII). HII's Newport News Shipbuilding (HII-NNS), the nation's only shipyard able to build nuclear-powered aircraft carriers, employs tens of thousands of skilled workers and earns billions of dollars in annual revenue, giving the company clear motivation to ensure the Navy retains its carrier-centric force structure.[11]

At HII, retired flag officers provide valuable knowledge, experience, and leadership that is beneficial for both the company and the Navy, and their employment is perfectly legal and ethical. However, their strong connections to active-duty leadership and Congress give them the ability to influence the Navy's force structure decisions, and their employment gives them an incentive to exert that influence. Consider some of the flag officers who served in key positions in the Navy that affected carrier construction at HII-NNS who have since retired and gone to work for HII:

- In January 2013 the four-star admiral serving as the director of Naval Reactors retired. After working at various other defense contractors, the retired admiral was hired by HII in 2017 to serve on its board of directors. In 2020 HII selected him to serve as chairman of the board of directors, where he relieved another retired four-star admiral.[12] Naval Reactors ensures safe and reliable operation of nuclear propulsion plants, including those on carriers built at HII-NNS.
- In June 2020, the three-star admiral serving as the commander of NAVSEA retired. Five months later, HII hired him as a vice president.[13] The Naval Sea Systems Command (NAVSEA) oversees public and private shipyards, including carrier construction at HII-NNS.
- In October 2020 the three-star admiral serving as COMNAVAIRFOR retired. Four months later, HII hired him as a corporate vice president, where he relieved another retired three-star admiral.[14] Commander, Naval Air Forces (COMNAVAIRFOR) is responsible for manning, training, and equipping naval aviation forces including aircraft carriers.
- In July 2019 the two-star admiral leading OLA retired. After working at another defense contractor, he was hired by HII in January 2021 as a corporate vice president.[15] The Office of Legislative Affairs (OLA) is responsible for most of the Navy's coordination with Congress, including discussions on force structure and shipyard performance.

So we see that some of the officers who were recently overseeing carrier design, operation, and financing have retired and gone to work for the sole producer of those carriers. Cancelling the *Ford* class would affect thousands of companies, many of which would oppose such a move. When just one of those companies has multiple highly respected retired three- and four-star admirals working for it in key executive roles, it makes it unlikely that the *Ford* class could be cancelled without a fight.

The Government

Even if the military and defense industry were to agree to stop building aircraft carriers, political realities may resist the decision.

"Political engineering" is the practice of sourcing parts and services from as many states and congressional districts as possible to ensure continued production, and in this regard the design of the *Ford* class is impressive. As of 2019 carrier construction and maintenance involved companies from 46 states and 293 congressional districts, meaning nearly every senator and two-thirds of representatives have strong incentives—constituent jobs—to keep building aircraft carriers.[16]

In 2019 the Navy proposed not refueling USS *Harry S. Truman* (CVN 75) to retire her early and redirect the funds to other projects, including unmanned aerial and surface vehicles. Capitol Hill responses ranged from "ridiculous idea," to "highly, highly skeptical," and "zero chance" the House Armed Services Subcommittee on Seapower and Projection Forces would take up the proposal.[17] Faced with this pushback, Vice President Mike Pence went on board *Truman* to announce that the Navy would refuel the carrier and continue to operate it, thus reversing the administration's own plan.[18] The attempt to not refuel *Truman* met the same fate as President Barack Obama's attempt to not refuel *Abraham Lincoln* (CVN 72) and *George Washington* (CVN 73).

Even if these administrations had not abandoned efforts to retire a carrier early, they would be faced with an additional hurdle—the law. Title 10 of the United States Code states, "The naval combat forces of the Navy shall include not less than eleven operational aircraft carriers," meaning the Navy's leaders cannot even propose a revamped fleet structure without congressional action.[19]

The Navy could not convince Congress to retire a single carrier early. Obtaining congressional approval to stop building large carriers entirely seems dubious. Any effort to alter the fleet's structure will be impossible without buy-in from all three components—military, industrial, and political, a group that former vice chairman of the Joint Chiefs of Staff Adm. James Winnefeld called the "iron triangle."[20]

Christian Brose, former staff director for the Senate Armed Services Committee, witnessed firsthand how difficult it is to unify those three entities to achieve meaningful change, in part because they all have their own objectives and incentives. Examining the current revolution in military affairs in his book *Kill Chain*, he wrote,

> This disruption will impact the fortunes of major companies that build legacy military equipment. It will call into question the work now being done by hundreds of thousands of Americans in uniform. And it will threaten the livelihoods of potentially millions of Americans who have long derived a sense of pride and dignity from the exquisite work they perform on behalf of the nation's defense. As the numbers of those disaffected and displaced by emerging technologies grow, they will find ample opportunity in America's defense budget process to resist the pace of disruption. We should sympathize with their reasons for doing so, even as we recognize the challenge it poses.[21]

INDIVIDUAL INERTIA

Overcoming human nature may be even more difficult than convincing key institutions. As Adm. William Sims said at the Naval War College's 1921 graduation, "Ever since men first began to use weapons to fight each other, military men have been reproached for excessive conservatism, a polite term often intended to imply a dangerous class reluctance to accept new ideas."[22]

The natural human tendency to resist change can lead to conclusions people might not otherwise make. For example, an aircraft carrier can certainly absorb a lot of punishment without sinking, yet a retired four-star admiral and former commander of U.S. Pacific Command went so far as to argue that if a submarine attacked an aircraft carrier, the ship's "hull structure [would be] itself a great defense against torpedoes."[23]

In the face of anti-ship missiles, many assume that escorts' defensive missiles will provide an umbrella of protection. But Standard Missile interceptors must track and hit targets with a cross-sectional area of approximately one square meter, traveling hundreds or thousands of miles per hour in three dimensions, outside the atmosphere in some cases.[24] By comparison, those incoming missiles aim at a target with a cross-sectional area of more than 20,000 square meters traveling at approximately 30 knots in only two dimensions.[25] The argument that Aegis missiles will succeed every time, but anti-ship missiles will fail every time is evidence of the natural tendency to inflate one's own capabilities while diminishing those of foes.

Much of the individual inertia resisting a shift away from aircraft carriers has focused on dismissing Chinese ASBMs. A former CNO stated without evidence, "The precise angle of impact is essential for the effectiveness of the ASBM warhead."[26] In another article, the admiral serving as the CNO's director of air warfare suggested that ASBMs would not be a major threat because during the missile's final 20 to 30 seconds of flight, the carrier could move 300 to 350 meters, which the missile could not account for and would therefore miss.[27] However, navies have been hitting moving targets at sea for centuries. While an ASBM is an immense technical challenge, the U.S. Navy cannot hope that the Chinese have not designed an operational terminal homing system. Additionally, *Ford*-class carriers are 332 meters long, so even without final course corrections, the ASBM still has a reasonable chance of hitting the carrier by the admiral's own math.[28]

One hundred years ago Secretary of War Newton D. Baker, when hearing Gen. Billy Mitchell's claim of being able to sink a battleship with airplanes, proclaimed, "That idea is so damned nonsensical and impossible that I'm willing to stand on the bridge of a battleship while that nitwit tries to hit it from the air."[29] Would anyone offer to stand on the bridge of an aircraft carrier today while the Chinese fire an ASBM at it?

In any case, changing the fleet's structure ultimately will not be a result of carrier vulnerability; carriers have always been vulnerable. Rather, change will come because the ability to launch an overwhelming strike against land or sea targets will wane compared to other options, and those better options will render moot the present compromise between the carrier's vulnerabilities and its awesome capabilities. As Fleet Adm. Chester Nimitz said after World War II in response to critics who questioned the carrier's survivability in the age of nuclear weapons, "Vulnerability of surface craft to atomic bombing does not necessarily mean that they have become obsolete. What determines the obsolescence of a weapon is not the fact that it can be destroyed, but that it can be replaced by another weapon that performs its functions more effectively."[30]

Yet the search for those more effective platforms and weapons, too, offers evidence of the natural tendency to dismiss the new in favor of the familiar. In a 2007 article advocating for aircraft carriers, a retired four-star admiral and former Vice Chief of Naval Operations, writing with two defense consultants, argued that cruise and conventional ballistic missiles "may have a role in certain campaigns, but it's likely to be a niche role."[31] No competent army would fight in the Bronze Age without bronze weapons, and no competent navy would fight in the Missile Age without missiles. The U.S. Navy cannot relegate missiles to a "niche" role and hope to win. One retired rear admiral took a scornful stance against critics of carriers, writing that defunding carrier strike groups was an idea coming from "left-leaning and libertarian think tanks as well as pundits of various stripes." He argued that without carrier forces, "The only alternative is to build, arm, man, and maintain foreign bases around the world forever." Aircraft carriers are certainly part of the U.S. Navy's future, but they are far from the only viable option. To dismiss other fleet structures as "absolute lunacy" will not help the Navy retain the dominance it has worked so hard to attain.[32]

OVERCOMING INERTIA

There are many valid reasons to keep building carriers and many deep thinkers have skillfully articulated them.[33] Admiral Sims, in the same speech in which he criticized the "dangerous class reluctance to accept new ideas," noted: "If, in general, such controversies have been based upon honest differences of opinion, sometimes strongly influenced by natural conservatism, still they were not free from the influences of our fallible human nature."[34]

Nonetheless, Admiral Sims also noted the need to consider change. He said the Navy's reluctance to adopt new ideas, weapons, and methods was "not the result of a lack of intelligence or patriotic interest, but was due chiefly to the long period during which our country was relatively free from foreign entanglements, and, consequently, when we so lacked the pressure of the probability of war . . . that we naturally thought we could afford to let other navies experiment with . . . new designs and weapons before we adopted them. We can no longer safely do so."[35]

The United States and the Navy face a similar challenge today. Whenever the Navy determines that fewer aircraft carriers are required, the speed of modern competition will demand that decision be enacted quickly. To overcome the aircraft carrier's institutional and individual inertia, four forces can move the Navy in a new direction. The best way to evolve is a combination of powerful leadership from the senior ranks and innovation from the junior ranks, working together to earn strong industrial and political support.

Top-Down Force

Most important will be a decision by Navy leaders to move beyond the carrier-centric fleet. The shift will involve a great deal more than just building a different kind of ship. With approximately 46 percent of all naval personnel serving on or supporting carriers, the impact will be tremendous, with effects on recruitment, training, weapons, tactics, maintenance, and supply.[36] Only the Navy's flag officers have the knowledge, skill, and resolve necessary to understand those impacts and lead the service through such a tumultuous transition. As Admiral Winnefeld said, for the U.S. Navy to transform its fleet structure, "The real leader has to be a noisy, impatient, creative, courageous, and insistent military leadership."[37]

Despite stumbles, the Navy's leaders historically have had great success overcoming inertia to implement change. Adm. Hyman Rickover spearheaded the development of the nuclear Navy, Adm. Elmo Zumwalt drove difficult personnel reforms, and the interwar Navy made incredible strides developing naval aviation, despite the myth of stodgy old "battleship admirals."

More recently, when Gen. David Berger was appointed commandant of the Marine Corps, he issued strategic guidance with fundamental changes to better prepare for war with China.[38] He waived the once sacrosanct requirement to have thirty-eight amphibious ships supporting insertion of two Marine expeditionary brigades, shifted the Marine Corps' tanks to the U.S. Army, and pushed infantry units to be smaller and more agile. His efforts to modernize the Marine Corps have met fierce resistance. A group of prominent retired Marines joined forces to publicly oppose General Berger's efforts while privately influencing Congress to do the same. That group includes every living commandant, a former secretary of defense, two former chairmen of the Joints Chiefs of Staff, and a former White House chief of staff, as well as approximately two dozen other generals, with many of them meeting daily to plan their way forward.[39]

When General Berger met with members of the group and refused to alter his progressive plan, a former Secretary of the Navy warned in the *Wall Street Journal* that "the gloves have now come off."[40] Despite some of those retired officers supporting the same modernization efforts while they were on active duty and being aware of the significant research and wargaming that went into the new force design, they now argue for a pause on the Marine Corp's modernization efforts for more "study and analysis," likely in an attempt to sentence those initiatives to death by committee.[41]

Just as General Berger must overcome significant inertia with retired general officers, he must do the same with the defense industry. One defense analyst went so far as to predict that the commandant's guidance threatens the "demise of the Marine Corps."[42] Despite that resistance, the Marine Corps continues to successfully implement General Berger's changes in its march toward modernization, a testament to the incredible results a determined active duty general or admiral can achieve.

Yet even the Marine Corps' modernization efforts would not be as tectonic a shift as the Navy forsaking nuclear-powered aircraft carriers. Designing such a new fleet structure would be an endeavor that has not been undertaken during the lives of any admiral or master chief. Fortunately, the Navy is not sitting idly by as new technologies and competitors arise but is already evolving its force structure with several exciting initiatives. In late 2020 CNO Adm. Michael Gilday signed out the "Project Overmatch" and "Novel Force" initiatives, concluding that the "Navy's ability to establish and sustain sea control in the future is at risk." He went on to suggest that one solution is a distributed fleet of manned and unmanned platforms exploiting AI/ML.[43] That same year, the Department of Defense completed the Future Naval Force Study (FNFS), an extensive analytical review of the Navy's structure. The FNFS, produced by the Office of the Secretary of Defense, Joint Chiefs of Staff, and the Navy, concluded that the fleet needed to shift to a larger, more distributed structure that makes better use of unmanned systems.[44] Additionally, the Navy has been advancing new hardware that will enable these initiatives, such as the *Constellation* class of frigates and the Conventional Prompt Strike hypersonic weapon.[45] Finally, in 2022 Admiral Gilday published his "Navigation Plan 2022," which identified six force design imperatives: distance, deception, defense, distribution, delivery, and decision advantage.[46]

Despite the promise of these efforts, the Navy will not be able to fully capitalize on its potential while continuing to build and rely on large nuclear-powered aircraft carriers. Although the FNFS envisions a distributed fleet of smaller platforms, the resulting Thirty-Year Shipbuilding Plan called for the continued construction of *Ford*-class carriers, with the U.S. Navy maintaining essentially the same number of CVNs in 2050 as it has in 2020.[47] Similarly, despite identifying force design imperatives such as deception and distribution, the CNO's Navigation Plan 2022 stated the Navy of 2045 should still have twelve nuclear-powered aircraft carriers.[48] Those carriers simply consume too many resources, precluding a larger fleet without a massive—and therefore unlikely—budget increase. More importantly, the Navy cannot fully benefit from the principles of Project Overmatch, Novel Force, the FNFS, and Navigation Plans—including the use of more numerous, distributed platforms with unmanned systems, missiles, and AI/ML—when the centerpiece of that fleet remains the centralized carrier strike group of large expensive manned ships. It is antithetical to design a fleet of advanced, distributed, and networked ships led by ten to twelve carrier strike groups.

As the U.S. Navy continues its efforts to modernize and evolve its architecture, it should heed the words of Admiral Sir John "Jackie" Fisher of the Royal Navy, who said, "In approaching . . . ship design, the first essential is to divest our minds totally of the idea that a single type of ship as now built is necessary, or even advisable."[49]

Bottom-Up Force

In the quest to evolve, the Navy will benefit from the input of its junior officers and sailors. Innovative, passionate, and with immense technical knowledge, the officers and sailors in the fleet can challenge conventions and offer realistic solutions.

Junior officers and sailors have a long history of overcoming inertia to implement needed change. As a junior officer, Lt. Cdr. William Sims spearheaded gunnery improvements despite resistance, even debating then–captain Alfred Thayer Mahan on the benefits of all-large-caliber warships.[50] World War II submarine captains and crews worked to troubleshoot defects with their torpedoes, even as the Bureau of Ordnance denied any problem. More recently, Maj. Leo Spaeder wrote a powerful article urging the Marine Corps to define its identity; his words may have influenced the commandant's strategic guidance.[51]

In all those cases, the critics went beyond citing a problem to providing solutions. As today's officers and sailors "read, think, and write," they need to move past discussions of carrier vulnerability.[52] The Navy needs a fleet that is technically, financially, and operationally realistic.[53] Most important, this new fleet needs to meet Nimitz's standard by showing that the carrier-centric fleet "can be replaced by another weapon that performs its functions more effectively."[54]

Internal Force

A much less likely means of overcoming the carrier's inertia is a substantial budgetary shift. With each *Gerald R. Ford*-class carrier costing approximately $12 billion, a massive cut in the Navy's budget would force it to select cheaper platforms regardless of capability. However, barring the simultaneous collapse of Russia and China, leaving the U.S. Navy to sail uncontested as it did in the 1990s, it is unlikely Congress would slash the budget enough to affect carrier production.

External Force

The final, and least desirable, means of implementing fleet change is naval combat. The CSS *Virginia*'s 1862 attack on wooden Union ships spurred rapid ironclad production. The 1941 Japanese destruction of HMS *Prince of Wales* and HMS *Repulse* with airplanes alone reinforced the need for aircraft carriers. A Chinese missile attack that disabled USS *Gerald R. Ford* would provide terrible momentum to redesign fleet structure.

100 YEARS AND 100,000 TONS

With essentially no conventional naval combat in the past 70 years, and with the last credible attack on an aircraft carrier coming at the Battle of Okinawa in 1945, it is hard to know when the aircraft carrier will pass into obsolescence. Whenever that moment does arrive, it will behoove the U.S. Navy to act on its own terms. It will not like the alternative—as Admiral Sims said, "It may be stated in general terms that most arguments in favor of fundamentally new weapons have failed, except those that resulted in shedding the blood of the unbelievers; that defeat alone has been accepted as a final demonstration."[55]

In March 1920, the USS *Jupiter* entered Norfolk Navy Yard to be converted into the U.S. Navy's first aircraft carrier.[56] Over the next hundred years, aircraft carriers performed splendidly, transforming the Navy into the most effective naval force in history. Despite this illustrious service, the Navy will eventually—whether in hundred days or hundred years—leave the carrier in its wake.

When the service decides it is time to change, institutional and individual opposition will not make it easy. The Navy and its people must persevere to keep it the most powerful force afloat, relying on strong leadership from the top, innovative solutions from the ranks, and earning buy-in from Congress and industry. Whatever course it charts, the Navy will need a large rudder to overcome the aircraft carrier's 100,000 tons of inertia.

➜⌇⌇CHAPTER **9**

THE SENKAKUS WAR

There seems to be something wrong with our bloody ships today![1]
—Adm. Sir David Beatty, RN

Yokosuka, Japan, August 8, 2028

The hypothetical Senkakus War of 2028 features the U.S. Navy's carrier-centric fleet and proposed Flex Fleet pitted against the People's Liberation Army Navy and supporting forces.

USS *Gerald R. Ford*, accompanied by three *Arleigh Burke*–class destroyers and two *Constellation*-class frigates, set sail from Yokosuka, Japan. Despite its departure under cover of darkness and under emissions control, the strike group's stealthy activity was duly noted by three Chinese nationals ashore as well as another on a trawler near Yokohama and quickly reported to the Chinese authorities. The next morning, satellite imagery confirmed the ships' movement. Munitions that had been on the pier were also gone, and dormant social media accounts indicated that *Ford* had embarked its entire air wing.

The U.S. Seventh Fleet had ordered the *Ford* strike group to sea in response to the Chinese seizure of the Senkaku Islands, an island chain just northeast

216

of Taiwan, three days earlier. More than two hundred Chinese marines had waded ashore onto the tiny, uninhabited islands and defiantly planted the Chinese flag. That same day, Chinese dredging ships commenced work creating a port while transports unloaded heavy construction equipment to build an airfield. That night, the Chinese Communist Party (CCP) issued a statement falsely declaring that China was the rightful owner of the Senkakus. It announced that it would not militarize the islands, claiming it was only constructing an oceanographic research facility and services to support Chinese fishermen.

American and Japanese analysts quickly called foul, as it was clear that the CCP was developing another military outpost. The CCP was likely attempting to distract its populace from the 2027 Hong Kong revolution, which Communist forces had brutally crushed, as well as the subsequent Uighur riots and nationwide unrest. U.S. military leaders were concerned that the seizure of the Senkakus was both a precursor to an invasion of Taiwan and a test of Japanese and American resolve. By acting aggressively, but only seizing an uninhabited, rocky island chain of little value, the CCP calculated that it could make another advance toward hegemony in the western Pacific without inciting a full-scale war.[2]

The next morning, the Japanese destroyer JDS *Kongō* (DDG 173) arrived off the largest island in the Senkakus, Uotsuri Jima. *Kongō*, as well as a Japanese Maritime Self-Defense Force (JMSDF) P-8 aircraft, was tasked with investigating the night's events and providing information on the disposition of Chinese forces. In Tokyo and Washington, the Japanese prime minster and U.S. president watched the events unfold in real time as they struggled with the decision on whether to contest the Chinese actions.

As *Kongō* approached the amphibious vessels and dredgers, her crew energized a fire control radar to refine the contact picture, just as the 5-inch gun swung through its full arc of motion as part of a routine daily check. However, four miles off *Kongō*'s bow, the *Jiangkai*-class frigate *Yiyang* (FFG 548) confused these actions for indicators of an imminent attack. *Yiyang*'s crew, on edge and tasked with protecting the vulnerable amphibious vessels, went to battle stations and informed other nearby PLAN vessels of the supposed threat. When *Kongō*'s helicopter took off for a reconnaissance flight, a *Yiyang* watchstander saw a radar return breaking away from the Japanese ship and incorrectly called away a missile launch. Not waiting to investigate, the *Yiyang* captain ordered a counterstrike, launching two YJ-83 ASCMs. At such short range, and with the CIWS in standby, *Kongō* had little chance to defend herself. The missiles ripped through her hull, instantly killing forty-four officers and sailors and causing the ship to take on a heavy list.

The president immediately knew he would have to order U.S. forces to strike back. Although the islands were insignificant and thousands of miles away from the United States, he could not let this act of Chinese aggression go unanswered without serious implications for U.S. credibility overseas. If he did not act, other partners would question the value of an American alliance, other adversaries would be emboldened, and the U.S.-led world order could be called into question.[3] Furthermore, as he lagged in the polls for the impending election, the president needed an important event to change the focus of the race.

The morning following the *Kongō* attack, the president announced that the United States would honor its obligations to Japan and that Chinese forces needed to either immediately withdraw from the Senkakus or be expelled by force. He declared, "The United States of America will not appease the Chinese Communist Party with the land and blood of our allies." In orders to his commanders, the president ordered the destruction of those Chinese air and naval forces necessary to recapture the Senkakus. In an effort to limit the war, prevent its escalation into a nuclear conflict, and minimize global economic effects, he forbade attacks on mainland China, only authorizing strikes against air, sea, space, and cyber assets, as well as ground forces on island outposts.[4]

For the first time in eighty years, the U.S. Navy went to war.

THE SECOND BATTLE OF THE PHILIPPINE SEA

Despite initiating the conflict, the CCP never imagined that the United States would be willing to fight for the Senkakus. As a result, huge portions of the PLAN remained in port and unprepared for war, while many Chinese aircraft needed maintenance, fuel, and munitions before they would be combat ready. In the days following the president's announcement, an uneasy calm developed in the western Pacific, as both sides prepared their forces for conflict and waited to see if the other side was serious about fighting.

However, on August 14, with the most vulnerable U.S. forces withdrawn from bases near mainland China, and with most of the Seventh Fleet at sea, the president relaxed the rules of engagement to allow attacks on Chinese forces. The initial U.S. strike was brutal. Mostly executed by stealth forces, the focus was on blinding Chinese reconnaissance efforts. Aegis destroyers shot down Chinese satellites and U.S. Cyber Command launched devastating attacks on PLAN networks. The few U.S. submarines available in the area launched their Tomahawk missiles at radar and communication installations at places like Subi Reef, Fiery Cross Reef, and the temporary sites on the Senkakus, while U.S. Air Force stealth bombers devastated the airfield and ports on those islands. Meanwhile, USS *George H. W. Bush* (CVN 77) and her strike group raced from operations off of Iran to the Sunda Strait, planning to attack Chinese forces and outposts in the South China Sea.

When the PLAN tracked the *Ford* strike group departing Yokosuka, Chinese intelligence assessed it was likely heading east to regroup with the rest of the U.S. Navy. The *Ford* strike group was under orders to withdraw to Hawaii, but only after it had conducted its own raid on Chinese forces. Turning south after leaving Yokosuka, the strike group stayed outside the so-called first island chain, passing four hundred miles east of Okinawa. At sunrise on the 16th, *Ford* launched almost its entire air wing on a strike designed to clear the seas and skies around the Senkakus. Dozens of F-35s, vectored in by E-2 Hawkeyes, sank four amphibious vessels unloading reinforcements, as well as two Luyang III destroyers, four Jiangkai II frigates, and a Jiangdao corvette. They easily shot down the six J-11B Flankers in the area, withdrawing before additional fighters could arrive from the mainland. Surface-to-air missiles shot down four F-35s in the attack; the submarine USS *Cheyenne* (SSN 773) rescued two of the pilots that night in a daring rescue.

After the raid, PLAN commanders realized they had been wrong about the carrier's intentions—and that, despite their losses, they had a chance to sink the carrier and win the war that day. However, they had to find it first. With much of the PLAN's scouting and communication networks disabled or slowed down, they had little situational awareness outside the first island chain. Satellite imagery was intermittent, several KJ-500 reconnaissance aircraft had been shot down by U.S. Air Force fighters as they evacuated Okinawa, and the remaining surface warships in the area had not found anything.

Nevertheless, Chinese commanders pushed their forces forward in a desperate attempt to find and attack *Ford* before she could withdraw. Surface action groups sped east from the East China Sea and the South China Sea, while more aircraft took off in a desperate scramble. Yet, with *Ford* already seven hundred miles from the mainland and sprinting east, the strike group commander was confident he could continue to open the distance and escape to the open Pacific.

At dusk, *Ford* and her escorts sighted their seventeenth merchant of the day, a nondescript tanker headed west. However, once over the horizon, the tanker captain attempted to raise his contact in the PAFMM. He struggled for more than an hour to use his PAFMM satellite phone, but with most Chinese military networks disabled he was never able to break through. In desperation, he used his personal computer to email his contact the simple message, "Carrier headed east, 2429N, 13114E," as well as a picture he had taken with his cell phone.

The email got through and unleashed pandemonium within the Chinese command structure. Immediately, the People's Liberation Army Rocket Force (PLARF) launched twenty DF-21D anti-ship ballistic missiles over an area centered on the reported position. However, due to the age of the report and lack of mid-course updates, they were all aimed well short of the *Ford* and her escorts. One tracked and obliterated a passing container ship while the rest fell harmlessly into the sea. Simultaneously, the PLAN and PLAAF launched 150 unmanned aerial vehicles (UAVs) to comb the seas near the reported position.

The hours ticked by as Chinese commanders waited for the UAVs to arrive and search the area. PLAN warships continued to sprint east, and H-6 Badger bombers fueled and took off from bases near Ningbo, close to Shanghai. Meanwhile, refueled F-35s took to the sky and shot down dozens of the incoming UAVs. As they continued their sweeps three hours before sunrise, a single BZK-005 UAV leaked through the defensive screen and found the *Ford* continuing its dash to the east.[5] It radioed the position just seconds before an escort's missile shot it down, but the damage was done. All the remaining UAVs in the area homed in on the position and started transmitting near-continuous tracking updates.

The E-2D Advanced Hawkeye detected the H-6 bombers, J-11B Flankers, and surface warships altered their courses to close with *Ford*, capitalizing on the new tracking data. The E-2D vectored in two squadrons of refueled F-35s while the Aegis escorts unleashed a salvo of missiles at the incoming aircraft. Dozens of bombers and fighters went down, as well as two more Luyangs and a Jiangkai. Two U.S. submarines, USS *Alexandria* (SSN 757) and USS *John Warner* (SSN 785), raced after the warships, picking off a large Renhai cruiser and three Jiangkais. Yet the Chinese kept coming.

They attacked piecemeal, unleashing their missiles as soon as they were in range before turning to avoid the wrath of the U.S. fighters and escorts. First came the YJ-12 ASCMs from the H-6 bombers, which had not detected the carrier but used the UAVs' targeting data.[6] Only three of the bombers had survived the F-35s' attack, but together they launched more than a dozen missiles. At the same time, the PLARF launched another salvo of ASBMs. After the first attack, only twelve were immediately available, and the communications link that provided midcourse updates remained inoperable. Finally, the three remaining surface ships within range launched a salvo of YJ-18A, YJ-62, and YJ-83 ASCMs at the extreme end of their range—but still outside that of many U.S. missiles.[7]

SM-2s, SM-3s, and SM-6 interceptors took down thirty-four incoming ASCMs, and close-in defenses destroyed another eleven. However, the destroyers were unable to launch missiles against the DF-21Ds. The communication system that provided the off-hull targeting data that enabled a "blind launch" was jammed, and until the ships' SPY radar detected the incoming missiles there was nothing to shoot at. *Ford* and her escorts conducted another large course change to try to throw off the incoming threats, while still heading roughly east to escape the Philippine Sea.

Just before dawn, a single YJ-12A slammed into the stern of USS *Mustin* (DDG 89). The detonation wrecked the ship's propellers and rudder, ruptured the flight deck, and killed forty-seven crewmembers instantly. *Mustin*, which had been closest to Chinese forces to screen *Ford*, slowed and settled by the stern. Moments later, with power lost to the ship's defense systems, three more ASCMs appeared on the horizon. Two hit the destroyer and ignited additional fires. USS *Constellation* (FFG 62) swung around to take on what survivors she could find as *Mustin* quicky sank stern first. *Constellation* retrieved just seven sailors before the strike group commander ordered her to abandon the search and resume her defensive position. Just as *Constellation* swung about, a YJ-83 detonated on *Ford*'s flight deck, destroying three refueling F-35s and igniting an inferno near the bow.

As the DF-21Ds began to appear on the destroyers' radar screens, making their terminal descent, the escorts launched a salvo of SM-6 interceptors. To that point, four of the twelve ASBMs had mechanical failures, and two more were improperly aimed. The SM-6 missiles destroyed three of the incoming missiles, but just minutes after sunrise one of the warheads detected and homed in on USS *Wayne E. Meyer* (DDG 108). The warhead penetrated most of the ship before detonating and blowing the ship in two. The entire ship disappeared beneath the waves within minutes. The two remaining ASBMs hurtled into the sea around the strike group, missing their targets but generating incredible geysers of water.

On *Ford*, the strike group commander looked at his tactical plot in amazement. His force had sunk fourteen PLAN warships, shot down forty-nine aircraft, and intercepted fifty-one missiles, yet the attack raged on. He had lost fourteen F-35s, an E-2D, and could feel the heat from the fire on his own flight deck. More importantly, he had seen two of his destroyers annihilated in a matter of minutes. Receiving a report that interceptors had missed another incoming YJ-12A, he thought back to the Battle of Jutland, dryly saying to a nearby officer, "There seems to be something wrong with our bloody ships today!"[8] Moments later, the missile detonated in *Ford*'s superstructure, instantly killing the admiral, ravaging the bridge, and disabling dozens of radars and communication systems.

Two more ASCMs penetrated the defensive screen, blowing two holes in *Ford*'s hull on the port side and temporarily jamming the rudder. After those hits, the Chinese attack petered out. The surviving U.S. destroyer reported no additional inbound missiles, and the UAVs were dropping into the sea from gunfire or lack of fuel. More U.S. submarines had found the PLAN warships and were inflicting a heavy toll, while additional destroyers from Hawaii and Guam raced to reinforce the strike group's defenses. The Chinese, unsure of what damage they had inflicted, called off subsequent attacks and withdrew the remaining ships and aircraft.

In the immediate aftermath, it was still unclear who had won what the press was calling the "Second Battle of the Philippine Sea." Tactically, U.S. forces had sunk sixteen warships, dozens of aircraft, and devastated what meager infrastructure and troops were on the Senkakus. In exchange, it had lost fifteen F-35, six other aircraft, two destroyers, and suffered heavy damage to *Ford*, although the ship was never at risk of sinking. However, the Senkaku Islands remained in Chinese hands.

The next week the battle was decided in the court of public opinion. Despite the press being kept away from *Ford*'s return, a camera-equipped drone captured incredible footage of *Ford* as she entered Pearl Harbor for emergency repairs. The drone's footage, posted online to an anonymous account, showed the ravaged hull, still-smoldering flight deck, and wrecked superstructure. The images sent shockwaves throughout the nation and were on the front page of nearly every newspaper in the world, overshadowing any tactical success the U.S. Navy had achieved.

The political consequences were immediate. The president's poll numbers plummeted as the country grew enraged at Chinese aggression and shocked by what it viewed as the U.S. Navy's poor performance. When news leaked that the *Bush* strike group had suffered minor damage from a mine in the Sunda Strait and withdrawn, congressional leaders demanded hearings on the Navy's performance. Yet with *Ford* entering drydock, *Bush* withdrawing for repairs away from the reported location of Chinese submarines, and most other carriers tied up in shipyard for major overhauls or still in the Atlantic, it was unclear where the president could turn to strike back at the Chinese.

THE FIGHT FOR THE FIRST ISLAND CHAIN

The day before Chinese marines landed on the Senkakus, the U.S. Pacific Fleet completed a "Fleet Battle Problem" near Guam. The complex exercise was the largest test of the Navy's experimental fleet structure—dubbed the Flex Fleet—ever undertaken; it included the Navy's new Steel-class corvettes, Brass-class missile arsenal ships, and Bronze-class light carriers, linked by the fleet's new all-encompassing tactical network. Since the inception of the Flex Fleet concept, the Navy had made great progress advancing the tactics, platforms, weapons, and infrastructure needed for effective distributed maritime operations, enabled by work like the Project Overmatch and Novel Force initiatives.[9] Most leaders within the military, defense industry, and the government understood its value, although they had not allowed its development to impede the construction of nuclear-powered aircraft carriers or affect their role as the flagship of the U.S. Navy.

With those carriers temporarily unavailable, the Seventh Fleet turned to the Flex Fleet to keep pressure on the Chinese until the rest of the fleet could arrive. Upon receipt of those orders, the task force commander immediately realized this was her chance to demonstrate the combat power of this evolved fleet structure. She quickly set plans in motion to push the Chinese back inside the first island chain and pave the way for the liberation of the Senkakus.

She settled on a phased approach to roll back PLA forces, focused on winning the fight for battlespace awareness. First, she designated geographic zones in the western Pacific to control which U.S. forces operated where. The Green Zone encompassed areas where the U.S. had complete situational awareness, allowing friendly forces to operate with impunity. In the Yellow Zone, scouting efforts were contested, and U.S. forces needed to have stealth abilities, small signatures, operate under emissions control, or be shielded by satellite and cyberattacks to maneuver safely. Finally, only stealth forces—submarines, stealth aircraft, and unmanned platforms—were allowed in the Red Zone, where the U.S. had limited visibility while the Chinese had good reconnaissance coverage.

Next, she set her attack plan in motion, with a focus on blinding the Chinese outside the first island chain. Satellite and cyberattacks continued, as did stealth-platform strikes on key radar and communication sites on island outposts. Meanwhile, the United States declared a massive exclusion zone in the western Pacific, warning that any civilian ship operating there would be assumed to be part of the PAFMM and attacked immediately to prevent the transmission of targeting data. U.S. aircraft specifically sought out reconnaissance aircraft and UAVs for destruction.

These efforts paved the way for combined U.S. assets—the Navy and the Air Force—to push forward into the western Pacific. The submarines unleashed a deadly campaign on the PLAN, inflicting heavy losses within the first island chain. Steel-class corvettes pushed into the Yellow Zone with raids into the Red Zone, probing for PLAN units and attacking with missiles if the ships or their embarked UAVs gained contact. While they achieved some success, the best pairing was either a submarine, corvette, frigate, or UAV providing targeting data to a Brass-class arsenal ship. The arsenal ships, hundreds of miles away and operating under emissions control, launched hundreds of missiles at PLAN targets. The pairing was far from perfect—the tactical net often failed, the missiles could take longer than an hour to arrive necessitating risky midcourse updates, and the fleet's missile inventory was always dangerously low—but it was effective overall. The frontline submarines, ships, and aircraft proved to be valuable as shooters, but they were priceless as sensors feeding the Navy's tactical grid. Backing all of this, Bronze-class light carriers provided fleet air defense. Operating from the Green Zone with forays into the Yellow Zone, their F-35s solely focused on deterring or destroying Chinese air attacks, leaving most other missions to other platforms. Together, little by little, the fleet inexorably drove the Chinese forces back.

In early September, with casualties becoming unbearable, the PLAN withdrew to port, a fact the CCP shielded from the population. PLAN leadership shifted strategies by harboring their strength and forcing the United States and Japan to either accept Chinese control of the Senkakus or conduct a risky amphibious landing. Throughout the campaign, Chinese ASCMs had sunk fourteen corvettes, six frigates, and three destroyers, and a DF-21D had destroyed a Bronze-class light carrier. Despite the losses, the Flex Fleet had proved resilient and maintained pressure on the PLAN, sinking twenty-one frontline warships and seventeen submarines.

Much of the Navy leadership was amazed at what those corvettes, frigates, submarines, and arsenal ships, paired with the fleet's destroyers and light carriers, had accomplished. What had started as a means of merely maintaining pressure on the Chinese until the capital ships could arrive had evolved into a winning strategy. Despite this success, the Chinese could still claim victory as long as they held the Senkakus. As U.S. ships and aircraft paused to rearm and regroup, the U.S. commanders finalized their plans to bring the full force of the fleet to bear in a final decisive blow.

THE BATTLE OF UOTSURI JIMA

On October 27 U.S. forces commenced a two-pronged assault designed to end the war that day. First, the U.S. Navy and the Air Force teamed up for a massive strike on military forces around Ningbo. In the war's first attack on the Chinese mainland, U.S. submarines, arsenal ships, and destroyers launched more than 275 Tomahawk missiles at the Eastern Theater Navy headquarters, shipyard facilities, radar installations, and docked ships. Meanwhile, Air Force bombers joined almost one hundred aircraft from three Bronze light carriers, a repaired *George H. W. Bush*, and the newly arrived *Harry S. Truman*. The targets were carefully selected to restrict the damage to military installations in the area closest to the Senkakus, with no attacks on CCP facilities or sites outside military bases. Land-based fighters and air defense systems intercepted dozens of U.S. missiles and shot down twenty-two aircraft, but many of the strikes were successful and the attacks proved devastating.

Later that morning, 175 U.S. Marines crept ashore onto the largest island in the Senkakus, Uotsuri Jima. Disembarking from U.S. submarines that had clandestinely delivered them, the Marines quickly overran the few surprised Chinese troops that had survived the previous weeks' bombardments. Once the island was secure, additional Marines arrived via *Berger*-class light amphibious warships; not enough to mount a serious defense against a concentrated counterattack, but enough to show to the world that the islands had been wrested away from the Chinese.[10]

That evening, the U.S. president offered a ceasefire to the PRC. On one hand, he offered an olive branch: a chance to end the hostilities, a return to the status quo before the war began, and the loosening of economic sanctions. On the other hand, he threatened that if the fighting continued, he would authorize attacks on more Chinese sites, would target CCP facilities, and would ensure U.S. broadcasts informed the entire Chinese population on how poorly the war was going for the PLA. Realizing the threat of further unrest was real and seeking to protect its power, the CCP quietly agreed to the end of hostilities.

As the U.S. fleet sailed home, the Navy looked back at the short but vicious war. Incredibly, *Ford* had tactically won the Second Battle of the Philippine Sea, but nearly lost the war for the United States in the process. After sustaining significant damage and giving the world reason to question the U.S. Navy's dominance, the fleet would never be the same. With so few other carriers immediately available, and with the president so hesitant to risk them, they had done little for the remainder of the fighting.

Instead, a new fleet structure had filled the void and won the war. It had embraced the Age of the Missile, made heavy use of its network of distributed ships, and struck with diversified kill chains that the Chinese had no answer for. It had focused on the early detection of the enemy, realizing that whoever found the enemy first was likely to attack effectively first. It was not a perfect fleet structure and suffered numerous losses, but it had proved resilient enough to complete the task at hand.

The CNO set out to convince the rest of the Navy's leadership, the defense industry, and elected leaders of the need to continue the fleet's evolution. He incorporated their feedback and built a consensus while holding firm on his vision for the Navy's future. Key among those to convince was the president, who had ridden a wave of public support after the Senkakus War to win a second term in office. Using the political capital gained from his election victory, the president canceled the contract for the rest of the *Ford* class and approved the CNO's ambitious plan. The president trusted the Navy's leaders to determine the exact force structure that would best serve the nation going forward, confident it would no longer revolve around the large nuclear-powered aircraft carrier.

⤳Notes

INTRODUCTION

1. Worthington Chauncey Ford, *The Writings of George Washington* (New York: Knickerbocker Press, 1891), 407.
2. Nicholas A. Lambert, "What Is a Navy For?" U.S. Naval Institute *Proceedings*, April 2021, https://www.usni.org/magazines/proceedings/2021/april/what-navy.
3. U.S. Department of Defense, "Annual Report to Congress: Military and Security Developments Involving the People's Republic of China" (Washington, DC: Office of the Secretary of Defense, 2021), vi.
4. The "kill chain" is the series of steps a platform or platforms execute to destroy an enemy, including searching for and tracking the target and then deciding, executing, and assessing the attack.
5. Stuart E. Johnson and Vice Admiral Arthur K. Cebrowski, USN (Ret.), "Alternative Fleet Architecture Design," *Center for Technology and National Security Policy, National Defense University*, August 2005, 72.
6. Ward Carroll and Bill Hamblet, "ADM Winnefeld Warns Winter is Coming" with Adm. James Winnefeld, USN (Ret.), U.S. Naval Institute *Proceedings* podcast, episode 171, July 12, 2020, https://www.usni.org/magazines/proceedings/the-proceedings-podcast/proceedings-podcast-episode-171-adm-winnefeld-warns.
7. Admiral John Richardson and Lieutenant Ashley O'Keefe, USN, "Now Hear This—Read. Write. Fight." U.S. Naval Institute *Proceedings*, June 2016, https://www.usni.org/magazines/proceedings/2016/june/now-hear-read-write-fight.
8. Capt. Alfred Thayer Mahan, USN, *Lessons of the War with Spain* (London: Sampson Low, Marston, 1899), 10–11.

PART 1. OPPORTUNITIES

1. Adm. I. J. Galantin, USN (Ret.), *Submarine Admiral: From Battlewagons to Ballistic Missiles* (Chicago: University of Illinois Press, 1995), 252.

CHAPTER 1. OPPORTUNITIES IN FLEET DESIGN

1. Capt. Wayne P. Hughes, USN (Ret.), *Fleet Tactics and Coastal Combat*, 2nd ed. (Annapolis: Naval Institute Press, 2000), 27.
2. Stuart E. Johnson and Vice Adm. Arthur K. Cebrowski, USN (Ret.), "Alternative Fleet Architecture Design," *Center for Technology and National Security Policy, National Defense University*, August 2005, 4.
3. D. K. Brown, *Nelson to Vanguard: Warship Design and Development 1923–1945* (Barnsley, England: Seaforth Publishing, 2000), 39.
4. Rear Adm. I. J. Galantin, USN, "The Future of Nuclear-Powered Submarines," U.S. Naval Institute *Proceedings*, June 1958, https://www.usni.org/magazines/proceedings /1958/june/future-nuclear-powered-submarines.
5. Hughes, *Fleet Tactics*, 27.
6. Hughes, *Fleet Tactics*, 226.
7. Capt.Robert C. Rubel, USN (Ret.), "Count Ships Differently," U.S. Naval Institute *Proceedings*, June 2020, https://www.usni.org/magazines/proceedings/2020/june /count-ships-differently.
8. "U.S. Ship Force Levels," Naval History and Heritage Command, last modified November 17, 2017, accessed July 9, 2020, https://www.history.navy.mil/research /histories/ship-histories/us-ship-force-levels.html.
9. "U.S. Ship Force Levels," Naval History and Heritage Command, last modified November 17, 2017, accessed September 20, 2020, https://www.history.navy.mil /research/histories/ship-histories/us-ship-force-levels.html.
10. Sam LaGrone, "USS *Mason* Fired 3 Missiles to Defend From Yemen Cruise Missile Attacks," *USNI News*, October 11, 2016, https://news.usni.org/2016/10/11 /uss-mason-fired-3-missiles-to-defend-from-yemen-cruise-missiles-attack.
11. Max Hastings and Simon Jenkins. *The Battle for the Falklands* (New York: W. W. Norton, 1983), 155.
12. Dennis M. Gormley, Andrew S. Erickson, and Jingdong Yuan, "A Potent Vector: Assessing Chinese Cruise Missile Developments," *Joint Force Quarterly* 75, October 2014: 99.
13. Rear Adm. Grant Sharp, USN, "Formal Investigation into the Circumstances Surrounding the Attack on the USS Stark (FFG 31) on 17 May 1987" (Washington, DC: Office of Information, Navy Department, 1987), 14.
14. Vitaliy O. Pradun, "From Bottle Rockets to Lightning Bolts: China's Missile Revolution and Strategy Against U.S. Military Intervention" *Naval War College Review* (Spring 2011): 11.
15. John C. Schulte, *An Analysis of the Historical Effectiveness of Anti-Ship Cruise Missiles in Littoral Warfare* (Monterey, CA: Naval Postgraduate School, September 1994), 16–17.

16. Capt. Arnold Henderson, USN (Ret.), "Training Against the Navy's #1 Threat," U.S. Naval Institute *Proceedings*, September 1997, https://www.usni.org/magazines/proceedings/1997/september/training-against-navys-1-threat.

17. Rubel, "Count Ships Differently."

18. Cdr. Phillip E. Pournelle, USN, "The Deadly Future of Littoral Sea Control," U.S. Naval Institute *Proceedings*, July 2015, https://www.usni.org/magazines/proceedings/2015–07/deadly-future-littoral-sea-control.

19. Chief of Naval Operations, "A Design for Maritime Superiority: Version 2.0" (Washington, DC: Department of the Navy, December 2018), 8.

20. Rubel, "Count Ships Differently."

21. Hughes, *Fleet Tactics*, 212.

22. Lincoln Paine, *The Sea and Civilization: A Maritime History of the World* (New York: Vintage Books, 2013), 489.

23. William Tuohy, *America's Fighting Admirals: Winning the War at Sea in World War II* (St. Paul, MN: Zenith Press, 2007), 159.

24. Ronald D. Utt, *Ships of Oak, Guns of Iron: The War of 1812 and the Forging of the American Navy* (New York: Regnery History, 2012), 177.

25. Hughes, *Fleet Tactics*, 179.

26. Hughes, *Fleet Tactics*, 194.

27. Christian Brose, *The Kill Chain* (New York: Hachette Books, 2020), 151.

28. Samuel Eliot Morison. *History of United States Naval Operations in World War II. Volume 6, Breaking the Bismarcks Barrier* (Annapolis: Naval Institute Press, 1950), 215–16.

29. Department of the Navy, "2017 U.S. Navy Program Guide" (Washington, DC: Department of the Navy, 2017), 133–36.

30. Capt. Arthur H. Barber III, USN (Ret.), "Redesign the Fleet," U.S. Naval Institute *Proceedings*, January 2019, https://www.usni.org/magazines/proceedings/2019/january/redesign-fleet.

31. Megan Eckstein, "Navy Focused on Strengthening Networks to Support Unmanned Operations," *USNI News* (October 27, 2020), https://news.usni.org/2020/10/27/navy-focused-on-strengthening-networks-to-support-unmanned-operations.

32. Johnson and Cebrowski, "Alternative Fleet Architecture Design," vii.

33. Pournelle, "The Deadly Future of Littoral Sea Control."

34. Adm. John Harvey Jr., USN (Ret.), Capt. Wayne Hughes Jr., USN (Ret.), Capt. Jeffrey Kline, USN (Ret.), and Lt. Zachary Schwartz, USN, "Sustaining American Maritime Influence," U.S. Naval Institute *Proceedings*, September 2013, https://www.usni.org/magazines/proceedings/2013/september/sustaining-american-maritime-influence.

35. Brose, *The Kill Chain*, xxix.

36. Vice Adm. Thomas Rowden, Rear Admiral Peter Gumataotao, and Rear Admiral Peter Fanta, USN, "Distributed Lethality," U.S. Naval Institute *Proceedings*, January 2015, https://www.usni.org/magazines/proceedings/2015/january /distributed-lethality.

37. Eckstein, "Navy Focused on Strengthening Networks to Support Unmanned Operations."

38. Cdr. Phillip E. Pournelle, USN, "When the U.S. Navy Enters the next Super Bowl, Will It Play like the Denver Broncos?" *War on the Rocks*, January 30, 2015, http://warontherocks.com/2015/01/when-the-u-s-navy-enters-the-next-super -bowl-will-it-play-like-the-denver-broncos/.

39. Rowden, Gumataotao, and Fanta, "Distributed Lethality."

40. Barber, "Redesign the Fleet."

41. Vice Adm. Arthur K. Cebrowski, USN, and John H. Garstka, "Network-Centric Warfare—Its Origin and Future," U.S. Naval Institute *Proceedings* (January 1998), https://www.usni.org/magazines/proceedings/1998/january /network-centric-warfare-its-origin-and-future.

CHAPTER 2. THE FLEX FLEET

1. Megan Eckstein, "Pentagon Leaders Have Taken Lead in Crafting Future Fleet from Navy," *USNI News*, June 24, 2020, https://news.usni.org/2020/06/24 /pentagon-leaders-have-taken-lead-in-crafting-future-fleet-from-navy.

2. Norman Polmar, *Aircraft Carriers: A History of Carrier Aviation and Its Influence on World Events, Volume I: 1909–1945* (Washington, DC: Potomac Books, Inc., 2006), 3–5.

3. Trent Hone, *Learning War* (Annapolis: Naval Institute Press, 2018), 139.

4. Capt. Henry J. Hendrix, USN, and Lt. Col. J. Noel Williams, USMC (Ret.), "Twilight of the $UPERfluous Carrier," U.S. Naval Institute *Proceedings*, May 2011, https:// www.usni.org/magazines/proceedings/2011/may/twilight-uperfluous-carrier.

5. Dr. Jerry Hendrix and Commander Benjamin Armstrong, USN, "The Presence Problem: Naval Presence and National Strategy," *Center for a New American Security*, January 2016, 2.

6. James Morgan, *Theodore Roosevelt: The Boy and the Man* (New York: The Mac-Millan Company, 1907), 216.

7. James Risen, "U.S. Warns China on Taiwan, Sends Warships to Area," *Los Angeles Times*, March 11, 1996, https://www.latimes.com/archives/la-xpm-1996 -03-11-mn-45722-story.html.

8. "A Cooperative Strategy for 21st Century Seapower." (Washington, DC: U.S. Navy, U.S. Marine Corps, and U.S. Coast Guard, 2007), http://ise.gov/sites /default/files/Maritime_Strategy.pdf.

9. Seth Cropsey, Bryan G. McGrath, and Timothy A. Walton, "Sharpening the Spear: The Carrier, the Joint Force, and High-End Conflict," *Hudson Institute*, October 2015, 29.

10. Office of the Chief of Naval Operations, "Report to Congress on the Annual Long-Range Plan for Construction of Naval Vessels for Fiscal Year 2020." (Washington, DC: Department of the Navy, March 2019), 13.

11. Bryan Clark, Peter Haynes, Bryan McGrath, Craig Hooper, Jesse Sloman, and Timothy A. Walton, "Restoring American Seapower: A New Fleet Architecture for the United States Navy," *Center for Strategic and Budgetary Assessments*, January 2017, 112–13.

12. Capt. Wayne P. Hughes, USN (Ret.), "The New Navy Fighting Machine: A Study of the Connections Between Contemporary Policy, Strategy, Sea Power, Naval Operations, and the Composition of the United States Fleet," Naval Postgraduate School, August 2009, 19.

13. Adm. John Harvey Jr., USN (Ret.), Capt. Wayne Hughes Jr., USN (Ret.), Capt. Jeffrey Kline, USN (Ret.), and Lieutenant Zachary Schwartz, USN, "Sustaining American Maritime Influence," U.S. Naval Institute *Proceedings*, September 2013, https://www.usni.org/magazines/proceedings/2013/september /sustaining-american-maritime-influence.

14. Harvey, Hughes, Kline, and Schwartz, "Sustaining American Maritime Influence."

15. "MQ-8 Fire Scout," Naval Air Systems Command, accessed July 12, 2016, http://www .navair.navy.mil/index.cfm?fuseaction=home.display&key=8250AFBA -DF2B-4999-9EF3-0B0E46144D03.

16. Hughes, "The New Navy Fighting Machine," 19.

17. Steel-class corvette cost estimate derived from Congressional Budget Office (CBO) estimates of surface combatant cost per thousand tons of displacement, as well as previous studies. Per the CBO's October 2020 report "The Cost of the Navy's New Frigate," the average cost of the Navy's surface combatants as a function of tonnage ranges between about $200–$300 million per thousand tons of displacement. Smaller warships, such as *Oliver Hazard Perry* frigates and littoral combat ships, tend to be at the lower end of that range, and only the *Ticonderoga*-class of cruisers has significantly exceeded it. The Steel-class corvette is 600 tons, so using a conservative cost estimate of $250 million per thousand tons, each corvette should cost $150 million. To be even more conservative, this cost estimate is doubled to use a per-price ship of $300 million. In Captain Wayne Hughes' "New Navy Fighting Machine" study for the Naval Postgraduate School, he outlined a proposed fleet with the Sea Lance corvette, which served as the model for the Steel-class corvette. That study used an estimate of $100 million

per corvette, which when converted to 2019 dollars equates to about $120 million, still less than half the conservative $300 million corvette cost used for Flex Fleet finances. Assuming the actual cost would be less than that, the Flex Fleet could purchase increased numbers of corvettes.

18. Cdr. Phillip E. Pournelle, USN, "We Need a Balanced Fleet for Naval Supremacy," *Information Dissemination*, November 4, 2013, http://www.information dissemination.net/2013/11/we-need-balanced-fleet-for-naval.html.

19. "Navy Frigate (FFG[X]) Program: Background and Issues for Congress" (Washington, DC: Congressional Research Service, June 26, 2020), 5, 20.

20. "Navy Frigate (FFG[X]) Program: Background and Issues for Congress," 31.

21. "Navy Frigate (FFG[X]) Program: Background and Issues for Congress," 7, 31.

22. Office of the Chief of Naval Operations, "Report to Congress on the Annual Long-Range Plan for Construction of Naval Vessels for Fiscal Year 2020," 13.

23. Hughes, "The New Navy Fighting Machine," 48.

24. Capt. R. Robinson Harris, USN (Ret.), Andrew Kerr, Kenneth Adams, Christopher Abt, Michael Venn, and Col. T. X. Hammes, USMC (Ret.), "Converting Merchant Ships to Missile Ships for the Win," U.S. Naval Institute *Proceedings*, January 2019, https://www.usni.org/magazines/proceedings/2019/january /converting-merchant-ships-missile-ships-win.

25. "*Tomahawk* Cruise Missile: U.S. Navy Fact File," U.S. Navy, last modified August 14, 2014, accessed May 30, 2016, http://www.navy.mil/navydata/fact_display .asp?cid=2200&tid=1300&ct=2. "Conventional Prompt Global Strike and Long-Range Ballistic Missiles: Background and Issues" (Washington, DC: Congressional Research Service, February 14, 2020), 42.

26. Sydney J. Freedberg Jr., "Polmar's Navy: Trade LCS & Carriers for Frigates and Amphibs," *Breaking Defense*, December 18, 2015, http://breakingdefense .com/2015/12/polmars-navy-trade-lcs-carriers-for-frigates-amphibs/.

27. Capt. Sam J. Tangredi, USN (Ret.), "Breaking the Anti-Access Wall," U.S. Naval Institute *Proceedings*, May 2015, https://www.usni.org/magazines/proceedings /2015/may/breaking-anti-access-wall.

28. Norman Polmar with Richard R. Burgess, *The Naval Institute Guide to the Ships and Aircraft of the U.S. Fleet*, 19th ed. (Annapolis: Naval Institute Press, 2013), 69.

29. Congressional Budget Office, "An Analysis of the Navy's Fiscal Year 2020 Shipbuilding Plan" (Washington, DC: Congress of the United States, October 2019), 20.

30. Hughes, "The New Navy Fighting Machine," 48. Cost increased from $200 to $220 million to account for inflation.

31. Capt. Wayne P. Hughes Jr., USN (Ret.), "Single-Purpose Warships for the Littorals," U.S. Naval Institute *Proceedings*, June 2014, https://www.usni.org/magazines/proceedings/2014/june/single-purpose-warships-littorals.

32. National Research Council, *Responding to Capability Surprise: A Strategy for U.S. Forces* (Washington, DC: National Academies Press, 2013), 73. Cited by Hughes, "Single-Purpose Warships for the Littorals."

33. Hughes, "The New Navy Fighting Machine," 47.

34. Office of the Chief of Naval Operations, "Report to Congress on the Annual Long-Range Plan for Construction of Naval Vessels for Fiscal Year 2020," 13.

35. Congressional Budget Office, "An Analysis of the Navy's Fiscal Year 2020 Shipbuilding Plan," 3, 20.

36. Ronald O'Rourke, "Navy Ford (CVN-78) Class Aircraft Carrier Program: Background and Issues for Congress" (Washington, DC: Congressional Research Service, June 8, 2020), 5.

37. Norman Polmar, *Aircraft Carriers: A History of Carrier Aviation and Its Influence on World Events, Volume II: 1946–2006* (Washington, DC: Potomac Books, 2008), 200.

38. Bradley Martin and Michael E. McMahon, *Future Aircraft Carrier Options* (Santa Monica, CA: RAND, 2017), 2.

39. Capt. Robert C. Rubel, USN (Ret.), "The Future of the Future of Aircraft Carriers," *Naval War College Review*, Autumn 2011: 24–25.

40. Capt. Henry J. Hendrix, USN, "Buy Fords, Not Ferraris," U.S. Naval Institute *Proceedings*, April 2009, https://www.usni.org/magazines/proceedings/2009/april/buy-fords-not-ferraris.

41. Office of the Chief of Naval Operations, "Report to Congress on the Annual Long-Range Plan for Construction of Naval Vessels for Fiscal Year 2020," 13.

42. Congressional Budget Office, "An Analysis of the Navy's Fiscal Year 2020 Shipbuilding Plan," 3, 20.

43. Martin and McMahon, *Future Aircraft Carrier Options*, ix, xi, 33.

44. Rubel, "The Future of the Future of Aircraft Carriers," 21.

45. Martin and McMahon, *Future Aircraft Carrier Options*, 16.

46. Norman Friedman, "F-35B Expands Definition of 'Aircraft Carriers,'" U.S. Naval Institute *Proceedings*, February 2019, https://www.usni.org/magazines/proceedings/2019/february/f-35b-expands-definition-aircraft-carriers.

47. Hughes, "The New Navy Fighting Machine," 26.

48. Martin and McMahon, *Future Aircraft Carrier Options*, 54. Price adjusted for inflation from 2017 to 2019 dollars.

49. Congressional Budget Office, "An Analysis of the Navy's Fiscal Year 2020 Shipbuilding Plan," 3, 20.

50. Office of the Chief of Naval Operations, "Report to Congress on the Annual Long-Range Plan for Construction of Naval Vessels for Fiscal Year 2020," 13.

51. Megan Eckstein, "Pentagon Leaders Have Taken Lead in Crafting Future Fleet from Navy."

CHAPTER 3. THE FLEX FLEET VS. THE CARRIER-CENTRIC FLEET

1. Capt. Robert C. Rubel, USN (Ret.), "The Future of the Future of Aircraft Carriers," *Naval War College Review*, Autumn 2011, 26.

2. Rear Adm. I. J. Galantin, USN, "The Future of Nuclear-Powered Submarines," U.S. Naval Institute *Proceedings*, June 1958, https://www.usni.org/magazines/proceedings/1958/june/future-nuclear-powered-submarines.

3. Office of the Chief of Naval Operations, "Report to Congress on the Annual Long-Range Plan for Construction of Naval Vessels for Fiscal Year 2020" (Washington, DC: Department of the Navy, March 2019), 13.

4. Seth Cropsey, Bryan G. McGrath, and Timothy A. Walton, "Sharpening the Spear: The Carrier, the Joint Force, and High-End Conflict," *Hudson Institute*, October 2015, 29.

5. "Navy Force Structure and Shipbuilding Plans: Background and Issues for Congress" (Washington, DC: Congressional Research Service, July 28, 2020), 2.

6. Norman Polmar with Richard R. Burgess, *The Naval Institute Guide to the Ships and Aircraft of the U.S. Fleet*, 19th ed. (Annapolis: Naval Institute Press, 2013), 1.

7. Dr. Jerry Hendrix and Commander Benjamin Armstrong, USN, "The Presence Problem: Naval Presence and National Strategy," *Center for a New American Security*, January 2016, 1.

8. Congressional Budget Office, "An Analysis of the Navy's Fiscal Year 2020 Shipbuilding Plan" (Washington, DC: Congress of the United States, October 2019), 3.

9. Harlan Ullman, "There is Now Only One Path to 355 Ships," U.S. Naval Institute *Proceedings*, April 2020, https://www.usni.org/magazines/proceedings/2020/april/there-now-only-one-path-355-ships.

10. Adm. Robert J. Natter, USN (Ret.), and Adm. Samuel J. Locklear, USN (Ret.), "Former 4-Star Fleet Commanders: Don't Give Up on Carriers," *Defense News*, November 22, 2019, https://www.defensenews.com/naval/2019/11/22/former-4-star-fleet-commanders-dont-give-up-on-carriers/.

11. Capt. Henry J. Hendrix, USN (Ret.) (PhD), "At What Cost a Carrier?" *Center for a New American Security Disruptive Defense Papers Series*, March 2013, 5.

12. Andrew S. Erickson, *Chinese Anti-Ship Ballistic Missile Development: Drivers, Trajectories and Strategic Implications* (Washington, DC: Jamestown Foundation, 2013), 31.

13. Michael J. Mazarr, *Understanding Deterrence* (Santa Monica, CA: RAND Corporation, 2018), 1.

14. Hendrix and Armstrong, "The Presence Problem," 15.

15. Norman Friedman, "F-35B Expands Definition of 'Aircraft Carriers,'" U.S. Naval Institute *Proceedings*, February 2019, https://www.usni.org/magazines/proceedings/2019/february/f-35b-expands-definition-aircraft-carriers.

16. Capt. Wayne P. Hughes, USN (Ret.), "The New Navy Fighting Machine: A Study of the Connections Between Contemporary Policy, Strategy, Sea Power, Naval Operations, and the Composition of the United States Fleet," Naval Postgraduate School, August 2009, 33.

17. Vice Adm. Thomas Rowden, Rear Adm. Peter Gumataotao, and Rear Adm. Peter Fanta, USN, "Distributed Lethality," U.S. Naval Institute *Proceedings*, January 2015, http://www.usni.org/magazines/proceedings/2015–01/distributed-lethality.

18. Christian Brose, *The Kill Chain* (New York: Hachette Books, 2020), 151.

19. Max Hastings and Simon Jenkins. *The Battle for the Falklands* (New York: W. W. Norton, 1983), 228.

20. Commander Phillip E. Pournelle, USN, "We Need a Balanced Fleet for Naval Supremacy," *Information Dissemination*, November 4, 2013, http://www.informationdissemination.net/2013/11/we-need-balanced-fleet-for-naval.html.

21. Pournelle, "We Need a Balanced Fleet for Naval Supremacy."

22. Vice Adm. James R. Fitzgerald, USN (Ret.), "More than Submarine vs. Submarine," U.S. Naval Institute *Proceedings*, February 2013, https://www.usni.org/magazines/proceedings/2013/february/more-submarine-vs-submarine.

23. Bradley Martin and Michael E. McMahon, *Future Aircraft Carrier Options* (Santa Monica, CA: RAND, 2017), 16.

24. Rear Admiral Yedidia Ya'ari, Israeli Navy, "The Littoral Arena: A Word of Caution," *Naval War College Review*, Spring 1995, 11.

25. Sam LaGrone, "USS *Mason* Fired 3 Missiles to Defend from Yemen Cruise Missile Attacks," *USNI News*, October 11, 2016, https://news.usni.org/2016/10/11/uss-mason-fired-3-missiles-to-defend-from-yemen-cruise-missiles-attack.

26. John C. Schulte, *An Analysis of the Historical Effectiveness of Anti-Ship Cruise Missiles in Littoral Warfare* (Monterey, CA: Naval Postgraduate School, September 1994), 16–17.

27. Hughes, "The New Navy Fighting Machine," 46.

28. Stuart E. Johnson and Vice Adm. Arthur K. Cebrowski, USN (Ret.), "Alternative Fleet Architecture Design," *Center for Technology and National Security Policy, National Defense University*, August 2005, 45.

29. Martin and McMahon, *Future Aircraft Carrier Options*, 8.

30. Adm. James L. Holloway III, USN (Ret.), *Aircraft Carriers at War* (Annapolis: Naval Institute Press, 2007), 441.

31. "The New Face of Naval Strike Warfare" *RAND National Defense Research Institute Research Brief*, 2005, 1.

32. Lockheed Martin, "F-35 Weaponry," accessed August 21, 2020, https://www .f35.com/about/carrytheload/weaponry#:~:text=In%20stealth%20mode%2C%20 the%20F,battle%20to%20finish%20the%20fight. "Tomahawk Cruise Missile: U.S. Navy Fact File," U.S. Navy, last modified August 14, 2014, accessed May 30, 2016, http://www.navy.mil/navydata/fact_display.asp?cid=2200&tid=1300&ct=2.

33. Lockheed Martin, "F-35 Weaponry."

34. Martin and McMahon, *Future Aircraft Carrier Options*, 42–43.

35. Dr. Jerry Hendrix, "Retreat From Range: The Rise and Fall of Carrier Aviation," *Center for a New American Security*, October 2015, 58.

36. Robert Haddick, *Fire on the Water: China, America, and the Future of the Pacific* (Annapolis: Naval Institute Press, 2014), 183.

37. Hendrix, "Retreat from Range," 58–61.

38. Michael Fabey, "US Navy CNO Cites Benefits of 'Aviation Combatants' for Future Force," *Janes*, October 14, 2020, https://www.janes.com/defence-news /news-detail/us-navy-cno-cites-benefits-of-aviation-combatants-for-future-force.

39. Capt. Wayne Hughes, USN (Ret.), "Restore a Distributable Naval Air Force," U.S. Naval Institute *Proceedings*, April 2019, https://www.usni.org/magazines /proceedings/2019/april/restore-distributable-naval-air-force.

40. Hughes, "Restore a Distributable Naval Air Force."

41. "Navy Shipyards: Actions Needed to Address the Main Factors Causing Maintenance Delays for Aircraft Carriers and Submarines" (Washington, DC: United States Government Accountability Office, August 2020), 7.

42. "Navy Shipyards: Actions Needed to Address the Main Factors Causing Maintenance Delays for Aircraft Carriers and Submarines," 11.

43. Ronald O'Rourke, "Navy *Ford* (CVN-78) Class Aircraft Carrier Program: Background and Issues for Congress" (Washington, DC: Congressional Research Service, December 17, 2019), 4.

44. Rubel, "The Future of the Future of Aircraft Carriers," 18.

45. David Larter, "U.S. Navy Looks to Hire Thousands More Sailors as Service Finds Itself 9,000 Sailors Short at Sea," *Navy Times*, February 10, 2020, https:// www.navytimes.com/smr/federal-budget/2020/02/11/us-navy-looks-to-hire -thousands-more-sailors-as-service-finds-itself-9000-sailors-short-at-sea/.

46. Capt. Wayne Hughes Jr., USN (Ret.), "22 Questions for Streetfighter," U.S. Naval Institute *Proceedings*, February 2000, https://www.usni.org/magazines /proceedings/2000/february/22-questions-streetfighter.

47. Adm. Jonathan Greenert, USN, "Payloads Over Platforms: Charting a New Course," U.S. Naval Institute *Proceedings*, July 2012, https://www.usni.org/magazines/proceedings/2012/july/payloads-over-platforms-charting-new-course.

48. Norman Polmar and Edward Whitman, *Hunters and Killers: Volume II: Antisubmarine Warfare from 1943* (Annapolis: Naval Institute Press, 2016), 212.

49. Hughes, "The New Navy Fighting Machine," 48.

50. Cdr. Phillip E. Pournelle, USN, "The Rise of the Missile Carriers," U.S. Naval Institute *Proceedings*, May 2013, http://www.usni.org/magazines/proceedings/2013–05/rise-missile-carriers.

51. Greenert, "Payloads Over Platforms."

52. Galantin, "The Future of Nuclear-Powered Submarines."

PART 2. RISKS AND OPPORTUNITY COSTS

1. Adm. Stansfield Turner, USN (Ret.), "Is the U.S. Navy Being Marginalized?" *Naval War College Review*, Summer 2003, 103.

CHAPTER 4. LESSONS FROM AIR AND SURFACE WARFARE HISTORY

1. Adm. Sandy Woodward, *One Hundred Days: The Memoirs of the Falklands Battle Group Commander* (Annapolis: Naval Institute Press, 1992), xviii.

2. Captain Tameichi Hara, IJN, with Fred Saito and Roger Pineau, *Japanese Destroyer Captain: Pearl Harbor, Guadalcanal, Midway—as Seen Through Japanese Eyes* (Annapolis: Naval Institute Press, 1967), 261.

3. James Belote and William Belote, *Typhoon of Steel: The Battle for Okinawa* (New York: Harper & Row, 1970), 106–8.

4. Clay Blair Jr., *Silent Victory; The U.S. Submarine War Against Japan* (Annapolis: Naval Institute Press, 1975), 832.

5. Belote and Belote, *Typhoon of Steel*, 108–10.

6. Belote and Belote, *Typhoon of Steel*, 117.

7. Robert Lecki, *Okinawa: The Last Battle of World War II* (New York: Penguin Group, 1995), 91.

8. Bill Sloan, *The Ultimate Battle: Okinawa 1945—The Last Epic Struggle of World War II* (New York: Simon & Schuster, 2007), 231.

9. Carl Boyd and Akihiko Yoshida, *The Japanese Submarine Force and World War II* (Annapolis: Naval Institute Press, 1995), 174.

10. Office of Program Appraisal, "Lessons of the Falklands" (Washington, DC: Department of the Navy, February 1983), 23.

11. Office of Program Appraisal, "Lessons of the Falklands," 23.

12. Victor Flintham, *Air Wars and Aircraft: A Detailed Record of Air Combat, 1945 to the Present* (New York: Facts on File, 1990), 231.

13. Adm. James L. Holloway III, USN (Ret.), *Aircraft Carriers at War* (Annapolis: Naval Institute Press, 2007), 53.

14. Holloway, *Aircraft Carriers at War*, 441.

15. John Prados, *Vietnam: The History of an Unwinnable War, 1945–1975* (Lawrence: University Press of Kansas, 2009), 94.

16. "Sterett III (DLG-31)," Naval History and Heritage Command, last modified April 28, 2020, accessed July 5, 2021, https://www.history.navy.mil/research /histories/ship-histories/danfs/s/sterett-iii.html.

17. Flintham, *Air Wars and Aircraft*, 90.

18. Brian L. Davis, *Qaddafi, Terrorism, and the Origins of the U.S. Attack on Libya* (New York: Praeger Publishers, 1990), 104.

19. Jeffrey L. Levinson and Randy L. Edwards, *Missile Inbound: The Attack on the Stark in the Persian Gulf* (Annapolis: Naval Institute Press, 1997), 7.

20. Rear Adm. Grant Sharp, USN, "Formal Investigation into the Circumstances Surrounding the Attack on the USS Stark (FFG 31) on 17 May 1987" (Washington, DC: Office of Information, Navy Department, 1987), 16.

21. Sharp, "Formal Investigation into the Circumstances Surrounding the Attack on the USS Stark (FFG 31) on 17 May 1987," 14.

22. Farhang Rajaee, *The Iran Iraq War: The Politics of Aggression* (Gainesville University Press of Florida, 1993), 141.

23. Harold Lee Wise. "One Day of War," *Naval History Magazine*, March 2013, https:// www.usni.org/magazines/naval-history-magazine/2013/march/one-day-war.

24. Wise, "One Day of War."

25. Barrett Tillman, "Fear and Loathing in the Post-Naval Era," U.S. Naval Institute *Proceedings*, June 2009, https://www.usni.org/magazines/proceedings/2009 /june/fear-and-loathing-post-naval-era.

26. Edward J. Marolda and Robert J. Schneller Jr., *Shield and Sword: The United States Navy and the Persian Gulf War* (Washington, DC: Naval Historical Center, Department of the Navy, 1998), 231.

27. Pelham G. Boyer and Robert S. Wood, eds., *Strategic Transformation and Naval Power in the 21st Century* (Newport, RI: Naval War College Press, 1998), 311.

28. Capt. Wayne P. Hughes, USN (Ret.), *Fleet Tactics and Coastal Combat*, 2nd ed. (Annapolis: Naval Institute Press, 2000), 152.

29. Andrew Hind, "The Cruise Missile Comes of Age," *Naval History Magazine*, October 2008, https://www.usni.org/magazines/naval-history-magazine/2008 /october/cruise-missile-comes-age.

30. Flintham, *Air Wars and Aircraft*, 197; and Major General D. K. Palit, Vr C (Ret.), *The Lightning Campaign: The Indo-Pakistan War 1971* (New Delhi: Thomson Press [India] Limited, 1972), 149.

31. Abraham Rabinovich, "From 'Futuristic Whimsy' to Naval Reality," *Naval History Magazine*, June 2014, https://www.usni.org/magazines/naval-history-magazine/2014/june/futuristic-whimsy-naval-reality.

32. John C. Schulte, "An Analysis of the Historical Effectiveness of Anti-Ship Cruise Missiles in Littoral Warfare," master's thesis, Naval Postgraduate School, September 1994, 8.

33. Ronald O'Rourke, "The Tanker War," U.S. Naval Institute *Proceedings*, May 1988, http://www.usni.org/magazines/proceedings/1988–05/tanker-war.

34. Amos Harel and Avi Issacharoff, *34 Days: Israel, Hezbollah, and the War in Lebanon* (New York: Palgreave MacMillan, 2008), 101.

35. Jeremy Binnie, "Sinai Militants Attack Egypt Patrol Boat," *IHS Jane's 360*, July 19, 2015, http://www.janes.com/article/53070/sinai-militants-attack-egyptian-patrol-boat.

36. Sam LaGrone, "Warship *Moskva* was Blind to Ukrainian Missile Attack, Analysis Shows," *USNI News*, May 5, 2022, https://news.usni.org/2022/05/05/warship-moskva-was-blind-to-ukrainian-missile-attack-analysis-shows.

37. Schulte, *An Analysis of the Historical Effectiveness of Anti-Ship Cruise Missiles in Littoral Warfare*, 8.

38. Norman Polmar with Richard R. Burgess, *The Naval Institute Guide to the Ships and Aircraft of the U.S. Fleet*, 19th ed. (Annapolis: Naval Institute Press, 2013), 3.

39. Daniel K. Gibran. *The Falklands War: Britain Versus the Past in the South Atlantic* (London: McFarland, 1998), 29.

40. Richard C. Thornton, *The Falklands Sting* (Washington: Brassey's, 1998), 130.

41. Gibran, *The Falklands War*, 74.

42. Lawrence Freedman, *The Official History of the Falklands Campaign. Volume II, War and Diplomacy* (London: Routledge, Taylor & Francis Group, 2005), 439–41.

43. Office of Program Appraisal, "Lessons of the Falklands," 2.

44. Freedman, *The Official History of the Falklands Campaign. Volume II*, 75.

45. Gordon Smith, *Battles of the Falklands War* (London: Ian Allen, 1989), 31.

46. Max Hastings and Simon Jenkins, *The Battle for the Falklands* (New York: W. W. Norton, 1983), 132.

47. Hastings and Jenkins, *The Battle for the Falklands*, 149.

48. Smith, *Battles of the Falklands War*, 61.

49. Hastings and Jenkins, *The Battle for the Falklands*, 155.

50. Smith, *Battles of the Falklands War*, 91.

51. Hastings and Jenkins, *The Battle for the Falklands*, 226.

52. Freedman, *The Official History of the Falklands Campaign. Volume II*, 487.

53. Freedman, *The Official History of the Falklands Campaign. Volume II*, 623.

54. Freedman, *The Official History of the Falklands Campaign. Volume II*, 215.

55. Commander Jorge Luis Colombo, ARA, "Falkland Operations I: 'Super Etendard' Naval Aircraft Operations during the Malvinas War," *Naval War College Review* (May/June 1984): 19.

56. Freedman, *The Official History of the Falklands Campaign. Volume II,* 256.

57. Office of Naval Intelligence, "The PLA Navy: New Capabilities and Missions for the 21st Century" (Washington, DC: Office of Naval Intelligence, 2015), 19, 22.

58. Andrew S. Erickson and Conor M. Kennedy, "China's Maritime Militia," Center for Naval Analyses, 7 March 2016, www.cna.org/cna_files/pdf/Chinas-Maritime -Militia.pdf.

59. Office of Program Appraisal, "Lessons of the Falklands," 32.

60. Hastings and Jenkins, *The Battle for the Falklands,* 102.

61. Dennis M. Gormley, Andrew S. Erickson, and Jingdong Yuan, *A Low-Visibility Force Multiplier: Assessing China's Cruise Missile Ambitions* (Washington, DC: National Defense University Press, 2014), 16.

62. Hughes, *Fleet Tactics and Coastal Combat,* 147–48.

63. Woodward, *One Hundred Days,* xviii.

64. Hastings and Jenkins, *The Battle for the Falklands,* 228.

65. Office of Program Appraisal, "Lessons of the Falklands," 6.

66. Freedman, *The Official History of the Falklands Campaign. Volume II,* 779.

67. Freedman, *The Official History of the Falklands Campaign. Volume II,* 779.

68. Woodward, *One Hundred Days,* 18–19.

69. Woodward, *One Hundred Days,* xviii.

70. Adm. Harry D. Train II, USN (Ret.), "An Analysis of the Falkland/Malvinas Islands Campaign," *Naval War College Review* (Winter 1988): 50.

71. Tillman, "Fear and Loathing in the Post-Naval Era."

72. Bradley Martin and Michael E. McMahon, *Future Aircraft Carrier Options* (Santa Monica, CA: RAND, 2017), 8.

73. Martin and McMahon, *Future Aircraft Carrier Options,* 42–43.

74. Capt. Robert C. Rubel, USN (Ret.), "Count Ships Differently," U.S. Naval Institute *Proceedings,* June 2020, https://www.usni.org/magazines/proceedings/2020/june /count-ships-differently.

75. Capt. Robert C. Rubel, USN (Ret.), "Connecting the Dots: Capital Ships, the Littoral, Command of the Sea, and the World Order," *Naval War College Review,* Autumn 2015, 49.

76. Rear Admiral Yedidia "Didi" Ya'ari, Israeli navy, "The Littoral Arena: A Word of Caution," *Naval War College Review,* Spring 1995, 19.

77. Rear Adm. Walter E. Carter Jr., USN, "Sea Power in the Precision-Missile Age," U.S. Naval Institute *Proceedings,* May 2014, https://www.usni.org/magazines /proceedings/2014/may/sea-power-precision-missile-age.

CHAPTER 5. AIR AND SURFACE WARFARE TODAY

1. Rear Admiral Yedida "Didi" Ya'ari, Israeli Navy, "The Littoral Arena: A Word of Caution," *Naval War College Review*, Spring 1995: 13. This quote is also cited by Capt. Robert C. Rubel, USN (Ret.), in his Autumn 2015 *Naval War College Review* article, "Connecting the Dots: Capital Ships, the Littoral, Command of the Sea, and the World Order."

2. U.S. Department of Defense, "Annual Report to Congress: Military and Security Developments Involving the People's Republic of China" (Washington, DC: Office of the Secretary of Defense, 2020), 44; Adm. Harry B. Harris Jr., USN, *U.S. Pacific Command Posture Testimony, Before the United States House of Representatives Armed Services Committee*, 115th Cong., April 26, 2017.

3. Defense Intelligence Agency, "Russia Military Power: Building a Military to Support Great Power Aspirations" (Washington, DC: Defense Intelligence Agency, 2017), 83.

4. Robert Haddick, *Fire on the Water: China, America, and the Future of the Pacific* (Annapolis: Naval Institute Press, 2014), 82.

5. Dennis M. Gormley, Andrew S. Erickson, and Jingdong Yuan, "A Potent Vector: Assessing Chinese Cruise Missile Developments," *Joint Force Quarterly 75*, October 2014: 103.

6. Andrew S. Erickson, *Chinese Anti-Ship Ballistic Missile Development: Drivers, Trajectories and Strategic Implications* (Washington, DC: The Jamestown Foundation, 2013), 31.

7. U.S. Department of Defense, "Annual Report to Congress: Military and Security Developments Involving the People's Republic of China" (Washington, DC: Office of the Secretary of Defense, 2020), 139.

8. Eric Heginbotham, et al., *The U.S.-China Military Scorecard* (Santa Monica, CA: RAND Corporation, 2015), 30.

9. U.S. Department of Defense. "Annual Report to Congress: Military and Security Developments Involving the People's Republic of China" (Washington, DC: Office of the Secretary of Defense, 2014), i.

10. Capt. James Fanell, USN, "Chinese Navy: Operational Challenge or Potential Partner?" (Panel at Naval Institute-AFCEA West 2014 Conference, San Diego, California, January 31, 2013).

11. U.S. Department of Defense, "Annual Report to Congress: Military and Security Developments Involving the People's Republic of China" (Washington, DC: Office of the Secretary of Defense, 2020), 44, 165.

12. Peter Dutton, Andrew S. Erickson, and Ryan Martinson, eds., *China's Near Seas Combat Capabilities* (Newport, RI: China Maritime Studies Institute, U.S. Naval War College, 2014), 122.

13. U.S. Department of Defense, "Annual Report to Congress: Military and Security Developments Involving the People's Republic of China" (Washington, DC: Office of the Secretary of Defense, 2020), 50.

14. U.S. Department of Defense, "Annual Report to Congress: Military and Security Developments Involving the People's Republic of China" (Washington, DC: Office of the Secretary of Defense, 2020), 52.

15. Dutton, Erickson, and Martinson, eds., *China's Near Seas Combat Capabilities*, 63.

16. U.S. Department of Defense, "Annual Report to Congress: Military and Security Developments Involving the People's Republic of China" (Washington, DC: Office of the Secretary of Defense, 2020), 50.

17. Toshi Yoshihara and James R. Holmes, *Red Star Over the Pacific: China's Rise and the Challenge to U.S. Maritime Strategy* (Annapolis: Naval Institute Press, 2010), 93.

18. Yoshihara and Holmes, *Red Star Over the Pacific*, 101–2.

19. Vitaliy O. Pradun, "From Bottle Rockets to Lightning Bolts: China's Missile Revolution and Strategy Against U.S. Military Intervention" *Naval War College Review*, Spring 2011: 13.

20. Dennis M. Gormley, Andrew S. Erickson, and Jingdong Yuan, *A Low-Visibility Force Multiplier: Assessing China's Cruise Missile Ambitions* (Washington, DC: National Defense University Press, 2014), 16.

21. "Harpoon Missile: U.S. Navy Fact File," U.S. Navy, last modified February 20, 2009, accessed May 24, 2015, http://www.navy.mil/navydata/fact_display .asp?cid=2200&tid=200&ct=2.

22. U.S. Department of Defense, "Annual Report to Congress: Military and Security Developments Involving the People's Republic of China" (Washington, DC: Office of the Secretary of Defense, 2015), 56.

23. Pradun, "From Bottle Rockets to Lightning Bolts," 13.

24. Eric Hagt and Matthew Durnin, "China's Antiship Ballistic Missile: Developments and Missing Links" *Naval War College Review*, Autumn 2009, 91.

25. Erickson, *Chinese Anti-Ship Ballistic Missile Development*, 1.

26. U.S. Department of Defense, "Annual Report to Congress: Military and Security Developments Involving the People's Republic of China" (Washington, DC: Office of the Secretary of Defense, 2020), 56.

27. Yoshihara and Holmes, *Red Star Over the Pacific*, 107.

28. Hagt and Durnin, "China's Antiship Ballistic Missile: Developments and Missing Links," 87.

29. Fanell, "Chinese Navy: Operational Challenge or Potential Partner?"

30. Capt. Wayne P. Hughes, USN (Ret.), *Fleet Tactics and Coastal Combat*, 2nd ed. (Annapolis: Naval Institute Press, 2000), 212.

31. Adm. Richard C. Macke, USN, "Now Hear This—Demise of the Aircraft Carrier? Hardly," U.S. Naval Institute *Proceedings*, October 2015, https://www.usni.org/magazines/proceedings/2015/october/now-hear-demise-aircraft-carrier-hardly.

32. Hughes, *Fleet Tactics*, 130–32.

33. Daniel G. Felger, *Engineering for Officer of the Deck* (Annapolis: Naval Institute Press, 1979), 102; Commander, Naval Surface Force, U.S. Pacific Fleet, "Naval Terminology," accessed January 10, 2021, https://www.public.navy.mil/surfor/Pages/Navy-Terminology.aspx.

34. Ronald H. Spector, *Eagle Against the Sun* (New York: Random House, 1985), 458.

35. U.S. Census Bureau, "World Population," accessed December 31, 2020, https://www.census.gov/data/tables/time-series/demo/international-programs/historical-est-worldpop.html.

36. Ian Urbina, "Stowaways and Crimes Aboard a Scofflaw Ship," *New York Times*, July 17, 2015, http://www.nytimes.com/2015/07/19/world/stowaway-crime-scofflaw-ship.html.

37. U.S. Department of Defense, "Annual Report to Congress: Military and Security Developments Involving the People's Republic of China" (Washington, DC: Office of the Secretary of Defense, 2020), 29.

38. Samuel Eliot Morison, *The Two Ocean War: A Short History of the United States Navy in the Second World War* (Boston: Little, Brown, 1963), 127–28.

39. Admiral Jonathan Greenert, USN, "Chief of Naval Operations Remarks" (Slade Gorton International Policy Center Luncheon, Seattle, Washington, September 24, 2013), accessed February 11, 2017, http://www.navy.mil/navydata/people/cno/Greenert/Speech/130924%20Slade%20Gorton%20International%20Policy%20Center%20Luncheon%20Seattle.pdf.

40. Lt. Cdr. Devere Crooks, USN, and Lt. Cdr. Mateo Robertaccio, USN, "The Face of Battle in the Information Age," U.S. Naval Institute *Proceedings*, July 2015, https://www.usni.org/magazines/proceedings/2015/july/face-battle-information-age.

41. Steven Norton and Clint Boulton, "Years of Tech Mismanagement Led to OPM Breach, Resignation of Chief" *Wall Street Journal*, July 10, 2015, http://blogs.wsj.com/cio/2015/07/10/years-of-tech-mismanagement-led-to-opm-breach-resignation-of-chief/; Ellen Nakashima and Craig Timberg, "Russian government hackers are behind a broad espionage campaign that has compromised U.S. agencies, including Treasury and Commerce," *Washington Post*, December 14, 2020, https://www.washingtonpost.com/national-security/russian-government-spies-are-behind-a-broad-hacking-campaign-that-has-breached-us-agencies-and-a-top-cyber-firm/2020/12/13/d5a53b88-3d7d-11eb-9453-fc36ba051781_story.html.

42. Union of Concerned Scientists, "UCS Satellite Database," last modified August 1, 2020, accessed January 10, 2021, http://www.ucsusa.org/nuclear_weapons

_and_global_security/solutions/space-weapons/ucs-satellite-database.html#
.VcTxBflVhBc.

43. Huntington Ingalls Industries, "Building a Giant: *Gerald R. Ford (CVN 78)*."

44. Admiral Jonathan Greenert, USN, "Navy 2025: Forward Warfighters," U.S. Naval Institute *Proceedings*, December 2011, https://www.usni.org/magazines /proceedings/2011/december/navy-2025-forward-warfighters.

45. Cdr. Phillip E. Pournelle, USN, "The Deadly Future of Littoral Sea Control," U.S. Naval Institute *Proceedings*, July 2015, https://www.usni.org/magazines /proceedings/2015/july/deadly-future-littoral-sea-control.

46. Cdr. Phillip E. Pournelle, USN, "When the U.S. Navy Enters the next Super Bowl, Will It Play like the Denver Broncos?" *War on the Rocks*, January 30, 2015, http:// warontherocks.com/2015/01/when-the-u-s-navy-enters-the-next-super-bowl-will -it-play-like-the-denver-broncos/.

47. Norman Polmar, "U.S. Navy—A Paradigm Shift," U.S. Naval Institute *Proceedings*, March 2012, https://www.usni.org/magazines/proceedings/2012/march/us -navy-paradigm-shift.

48. U.S. Department of Defense, "Annual Report to Congress: Military and Security Developments Involving the People's Republic of China" (Washington, DC: Office of the Secretary of Defense, 2020), 65.

49. U.S. Department of Defense, "Annual Report to Congress: Military and Security Developments Involving the People's Republic of China" (Washington, DC: Office of the Secretary of Defense, 2020), 51.

50. U.S. Department of Defense, "Annual Report to Congress: Military and Security Developments Involving the People's Republic of China" (Washington, DC: Office of the Secretary of Defense, 2020), 71.

51. Erickson, *Chinese Anti-Ship Ballistic Missile Development*, 20.

52. Marshall Hoyler, "China's 'Antiaccess' Ballistic Missiles and U.S. Active Defense," *Naval War College Review*, Autumn 2010: 98.

53. U.S. Department of Defense, "Annual Report to Congress: Military and Security Developments Involving the People's Republic of China" (Washington, DC: Office of the Secretary of Defense, 2015), 35.

54. Heginbotham, et al., *The U.S.-China Military Scorecard*, 130.

55. Harold Lee Wise. "One Day of War," *Naval History Magazine*, March 2013, https:// www.usni.org/magazines/naval-history-magazine/2013/march/one-day-war.

56. Norman Polmar with Richard R. Burgess, *The Naval Institute Guide to the Ships and Aircraft of the U.S. Fleet*, 19th ed. (Annapolis: Naval Institute Press, 2013), 349.

57. Yoshihara and Holmes, *Red Star Over the Pacific*, 99.

58. "China Vehicle Population Hits 240 Million as Smog Engulfs Cities" *Bloomberg Business*, January 31, 2013, https://www.bloomberg.com/news/articles/2013-02-01 /china-vehicle-population-hits-240-million-as-smog-engulfs-cities.

59. Hoyler, "China's 'Antiaccess' Ballistic Missiles and U.S. Active Defense," 98.

60. Alan J. Vick, Richard M. Moore, Bruce R. Pirnie, and John Stillion, *Aerospace Operations Against Elusive Ground Targets* (Santa Monica, CA: RAND, 2001), 86.

61. Alan J. Vick, Richard M. Moore, Bruce R. Pirnie and John Stillion, *Aerospace Operations Against Elusive Ground Targets* (Santa Monica, CA: RAND, 2001), 65–66.

62. Polmar with Burgess, *The Naval Institute Guide to the Ships and Aircraft of the U.S. Fleet*, 127.

63. Polmar with Burgess, *The Naval Institute Guide to the Ships and Aircraft of the U.S. Fleet*, 553.

64. Polmar with Burgess, *The Naval Institute Guide to the Ships and Aircraft of the U.S. Fleet*, 396.

65. Polmar with Burgess, *The Naval Institute Guide to the Ships and Aircraft of the U.S. Fleet*, 349.

66. Polmar with Burgess, *The Naval Institute Guide to the Ships and Aircraft of the U.S. Fleet*, 502.

67. "Standard Missile: U.S. Navy Fact File," U.S. Navy, last modified November 15, 2013, accessed August 9, 2015, http://www.navy.mil/navydata/fact_display .asp?cid=2200&tid=1200&ct=2.

68. Polmar with Burgess, *The Naval Institute Guide to the Ships and Aircraft of the U.S. Fleet*, 516.

69. U.S. Navy, "Evolved SeaSparrow Missile (ESSM) (RIM 162D): U.S. Navy Fact File," last modified November 19, 2013, accessed August 9, 2015, http://www .navy.mil/navydata/fact_display.asp?cid=2200&tid=950&ct=2.

70. Polmar with Burgess, *The Naval Institute Guide to the Ships and Aircraft of the U.S. Fleet*, 503.

71. U.S. Navy, "RIM 116 Rolling Airframe Missile (RAM): U.S. Navy Fact File," last modified November 19, 2013, accessed August 9, 2015, http://www.navy.mil /navydata/fact_display.asp?cid=2200&tid=950&ct=2.

72. Polmar with Burgess, *The Naval Institute Guide to the Ships and Aircraft of the U.S. Fleet*, 509.

73. Hughes, *Fleet Tactics*, 152.

74. U.S. Navy, "MK 53—Decoy Launching System (Nulka): U.S. Navy Fact File," last modified November 15, 2013, accessed August 9, 2015, http://www.navy.mil /navydata/fact_display.asp?cid=2100&tid=587&ct=2.

75. Polmar with Burgess, *The Naval Institute Guide to the Ships and Aircraft of the U.S. Fleet*, 493.

76. U.S. Navy, "F-14 Tomcat Fighter: U.S. Navy Fact File," last modified February 17, 2009, accessed May 24, 2015, http://www.navy.mil/navydata/fact_display.asp?cid

=1100&tid=1100&ct=1; Lockheed Martin, "F-35C Carrier Variant," accessed May 24, 2015, http://www.lockheedmartin.com/us/products/f35/f-35c-carrier-variant .html.

77. Director, Operational Test and Evaluation, "FY14 Annual Report" (Washington, DC: Office of the Director, Operational Test and Evaluation, January 2015), 171–72.

78. Capt. Wayne P. Hughes, USN (Ret.), "The New Navy Fighting Machine: A Study of the Connections Between Contemporary Policy, Strategy, Sea Power, Naval Operations, and the Composition of the United States Fleet," Naval Postgraduate School, August 2009, 46.

79. Hughes, *Fleet Tactics*, 290.

80. Rear Adm. Grant Sharp, USN. "Formal Investigation into the Circumstances Surrounding the Attack on the USS Stark (FFG 31) on 17 May 1987" (Washington, DC: Office of Information, Navy Department, 1987), 3.

81. Capt. William A. Hesser Jr., USN, "Command Investigation into the Target Drone Malfunction and Strike of the USS Chancellorsville (CG 62) on 16 November 2013" (Pearl Harbor, HI: Commander, U.S. Pacific Fleet, December 20, 2013), 33.

82. Cdr. Phillip E. Pournelle, "The Rise of the Missile Carriers," U.S. Naval Institute *Proceedings*, May 2013, https://www.usni.org/magazines/proceedings/2013/may /rise-missile-carriers.

83. Max Hastings and Simon Jenkins, *The Battle for the Falklands* (New York: W. W. Norton, 1983), 227.

84. Sharp, "Formal Investigation into the Circumstances Surrounding the Attack on the USS Stark (FFG 31) on 17 May 1987," 16.

85. Rear Adm. Walter E. Carter Jr., USN, "Sea Power in the Precision-Missile Age," U.S. Naval Institute *Proceedings*, May 2014, https://www.usni.org/magazines /proceedings/2014/may/sea-power-precision-missile-age.

86. Ward Carroll and Bill Hamblet, hosts, "ADM Winnefeld Warns Winter is Coming" with Admiral James Winnefeld, USN (Ret.). U.S. Naval Institute *Proceedings* podcast, episode 171, July 12, 2020, https://www.usni.org /magazines/proceedings/the-proceedings-podcast/proceedings-podcast -episode-171-adm-winnefeld-warns.

87. U.S. Department of Defense Missile Defense Agency, "A System of Elements," last modified April 16, 2015, accessed August 11, 2015, http://www.mda.mil/system /elements.html.

88. Rear Adm. Michael C. Manazir, USN. "Responsive and Relevant," U.S. Naval Institute *Proceedings*, February 2014, https://www.usni.org/magazines /proceedings/2014/february/responsive-and-relevant.

89. Ronald O'Rourke, "Navy Aegis Ballistic Missile Defense (BMD) Program: Background and Issues for Congress" (Washington, DC: Congressional Research Service, December 23, 2020), 5.

90. "Patriotisms," *Science*, 17 April 1992, http://www.sciencemag.org/content /256/5055/312.full.pdf?sid=8515261b-059b-4e73–93fd-0ea4b44daab4.

91. Lisbeth Gronlund, David C. Wright, George N. Lewis, and Philip E. Coyle III, "Technical Realities: An Analysis of the 2004 Deployment of a U.S. National Missile Defense System" (Cambridge, MA: Union of Concerned Scientists, May 2004), ix.

92. Ronald O'Rourke, "Navy Aegis Ballistic Missile Defense (BMD) Program: Background and Issues for Congress" (Washington, DC: Congressional Research Service, December 23, 2020), 34.

93. Ronald O'Rourke, "Navy Aegis Ballistic Missile Defense (BMD) Program: Background and Issues for Congress" (Washington, DC: Congressional Research Service, March 4, 2015), 15.

94. George N. Lewis and Theodore A. Postol, "A Flawed and Dangerous U.S. Missile Defense Plan," *Arms Control Today,* May 5, 2010, https://www.armscontrol.org /act/2010_05/Lewis-Postol.

95. Lewis and Postol, "A Flawed and Dangerous U.S. Missile Defense Plan."

96. Ronald O'Rourke, "Navy Aegis Ballistic Missile Defense (BMD) Program: Background and Issues for Congress" (Washington, DC: Congressional Research Service, March 4, 2015), 7.

97. U.S. Department of Defense, "Annual Report to Congress: Military Power of the People's Republic of China" (Washington, DC: Office of the Secretary of Defense, 2008), 56; U.S. Department of Defense, "Annual Report to Congress: Military and Security Developments Involving the People's Republic of China" (Washington, DC: Office of the Secretary of Defense, 2020), 59.

98. Adm. Harry B. Harris Jr., USN, *U.S. Pacific Command Posture Testimony, Before the United States House of Representatives Armed Services Committee*, 115th Cong., April 26, 2017.

99. Pradun, "From Bottle Rockets to Lightning Bolts," 29.

100. For this analysis, the speed and size of the comparable U.S. Pershing II missile is used for the DF-21D due to the unavailability of the Chinese missile's unknown exact parameters. "U.S. Navy Sees Chinese HGV as Part of Wider Threat," *Aviation Week and Space Technology,* Jaunary 27, 2014, http://aviationweek. com/awin/us-navy-sees-chinese-hgv-part-wider-threat; "Pershing II Weapon System Operator's Manual" (Washington, DC: Headquarters, Department of the Army, June 1, 1986), accessed August 13, 2015, http://pershingmissile.org /pershingdocuments/manuals/tm%209–1425–386–10–1.pdf.

101. Jonathan Broder, "What China's New Missiles Mean for the Future of the Aircraft Carrier" *Newsweek*, February 16, 2016, http://www.newsweek.com /china-dongfeng-21d-missile-us-aircraft-carrier-427063.

102. Cdr. John Patch, USN (Ret.), "Fortress at Sea? The Carrier Invulnerability Myth," U.S. Naval Institute *Proceedings*, January 2010, https://www.usni.org/magazines /proceedings/2010/january/fortress-sea-carrier-invulnerability-myth.

103. George C. Daughan, *1812: The Navy's War* (New York: Basic Books, 2011), 214.

104. Broder, "What China's New Missiles Mean for the Future of the Aircraft Carrier."

105. Admiral Stansfield Turner, USN (Ret.), "Is the U.S. Navy Being Marginalized?" *Naval War College Review*, Summer 2003: 102.

106. Emphasis in the original. Ya'ari, "The Littoral Arena: A Word of Caution," 13. This quote is also cited by Capt. Robert C. Rubel, USN (Ret.), in his Autumn 2015 *Naval War College Review* article, "Connecting the Dots: Capital Ships, the Littoral, Command of the Sea, and the World Order."

107. Emphasis in the original. Stuart E. Johnson and Vice Adm. Arthur K. Cebrowski, USN (Ret.), "Alternative Fleet Architecture Design," *Center for Technology and National Security Policy, National Defense University*, August 2005, 72.

108. Lt. David Adams, USN, "We Are Not Invincible," U.S. Naval Institute *Proceedings*, May 1997, https://www.usni.org/magazines/proceedings/1997/may /we-are-not-invincible.

CHAPTER 6. LESSONS FROM UNDERSEA WARFARE HISTORY

1. Peter Padfield, *War Beneath the Sea: Submarine Conflict During World War II* (New York: John Wiley & Son, 1995), 479.

2. Adm. Sandy Woodward, *One Hundred Days: The Memoirs of the Falklands Battle Group Commander* (Annapolis: Naval Institute Press, 1992), 158.

3. Max Hastings and Simon Jenkins. *The Battle for the Falklands* (New York: W. W. Norton, 1983), 164.

4. Woodward, *One Hundred Days*, 164.

5. Lieutenant Commander Jeff Vandenengel, USN, "Fighting Along a Knife Edge in the Falklands," U.S. Naval Institute *Proceedings*, December 2019, https://www.usni. org/magazines/proceedings/2019/december/fighting-along-knife-edge-falklands.

6. Alfred Thayer Mahan, *Armaments and Arbitration: The Place of Force in the International Relations of States* (New York: Harper and Brothers Publishers, 1912), 206.

7. Joshua J. Edwards and Captain Dennis M. Gallagher, USN, "Mine and Undersea Warfare for the Future," U.S. Naval Institute *Proceedings*, August 2014, https://www.usni.org/magazines/proceedings/2014/august/mine-and -undersea-warfare-future.

8. Norman Polmar and Edward Whitman, *Hunters and Killers: Volume II: Anti-Submarine Warfare from 1943* (Annapolis: Naval Institute Press, 2016), 36.

9. Naval History and Heritage Command, "United States Submarine Losses World War II: Introduction," last modified July 23, 2015, accessed November 28, 2015,

http://www.history.navy.mil/research/library/online-reading-room/title-list
-alphabetically/u/united-states-submarine-losses/introduction.html.

10. Commander Robert H. Smith Jr., USN, "The Submarine's Long Shadow," U.S. Naval Institute *Proceedings*, March 1966, https://www.usni.org/magazines /proceedings/1966/march/submarines-long-shadow.

11. Axel Niestlé, *German U-Boat Losses During World War II: Details of Destruction* (London: Frontline Books, 2014), 199.

12. Naval History and Heritage Command, "United States Submarine Losses World War II: Introduction," last modified July 23, 2015, accessed November 28, 2015, http://www.history.navy.mil/research/library/online-reading-room/title-list -alphabetically/u/united-states-submarine-losses/introduction.html.

13. Padfield, *War Beneath the Sea*, 331.

14. Padfield, *War Beneath the Sea*, 167.

15. Clay Blair Jr., *Silent Victory; The U.S. Submarine War Against Japan* (Annapolis: Naval Institute Press, 1975), 508.

16. Norman Polmar and Edward Whitman, *Hunters and Killers: Volume I: Anti-Submarine Warfare from 1776 to 1943* (Annapolis: Naval Institute Press, 2015), 159.

17. Samuel Eliot Morison, *The Two Ocean War: A Short History of the United States Navy in the Second World War* (Boston: Little, Brown, 1963), 494.

18. Morison, *The Two Ocean War*, 496.

19. Padfield, *War Beneath the Sea*, 83.

20. Norman Polmar, *Aircraft Carriers: A History of Carrier Aviation and Its Influence on World Events, Volume I: 1909–1945* (Washington, DC: Potomac Books, 2006), 134.

21. Polmar and Whitman, *Hunters and Killers: Volume II*, 58.

22. Padfield, *War Beneath the Sea*, 251–52.

23. Polmar, *Aircraft Carriers: A History of Carrier Aviation and Its Influence on World Events, Volume I: 1909–1945*, 395–96.

24. Padfield, *War Beneath the Sea*, 63, 177.

25. Capt. Wayne P. Hughes, USN (Ret.), *Fleet Tactics and Coastal Combat*, 2nd ed. (Annapolis: Naval Institute Press, 2000), 139. The USS *Yorktown*, although technically sunk by the Japanese submarine *I-168*, was heavily damaged at the time and capable of only 3 knots and so credit is given to aircraft.

26. Morison, *The Two Ocean War*, 493.

27. Padfield, *War Beneath the Sea*, 336.

28. Commo. Stephen Saunders, RN, *Jane's Fighting Ships 2011–2012*, 114th ed. (Alexandria, VA: Jane's Information Group, 2011), 135.

29. Thomas Parrish, *The Submarine: A History* (New York: Penguin Group, 2004), 451.

30. Morison, *The Two Ocean War*, 494; U.S. Navy, "Attack Submarines - SSN: U.S. Navy Fact File," last modified November 9, 2015, accessed January 2, 2016, http://www.navy.mil/navydata/fact_display.asp?cid=4100&ct=4&tid=100.

31. John Keegan, *The Price of Admiralty: The Evolution of Naval Warfare* (New York: Viking Penguin, 1989), 269.

32. Padfield, *War Beneath the Sea*, 401; General Dynamics, Electric Boat, "Our Submarines," accessed January 2, 2016, http://www.gdeb.com/about/oursubmarines/

33. Smith, "The Submarine's Long Shadow."

34. Capt. William Hoeft Jr., USN (Ret.), "Nobody Asked Me, But . . . In the Deep, Run Silent Again," U.S. Naval Institute *Proceedings*, January 2013, 12.

35. Dennis M. Gormley, Andrew S. Erickson, and Jingdong Yuan, "A Potent Vector: Assessing Chinese Cruise Missile Developments," *Joint Force Quarterly* 75, no. 4 (October 2014): 100–101.

36. Rear Adm. W. J. Holland, USN (Ret.), "SSN: The Queen of the Seas" *Naval War College Review*, Spring 1991: 117.

37. Keegan, *The Price of Admiralty*, 238.

38. Vice Adm. James R. Fitzgerald, USN (Ret.), "More than Submarine vs. Submarine," U.S. Naval Institute *Proceedings*, February 2013, https://www.usni.org/magazines/proceedings/2013/february/more-submarine-vs-submarine.

39. Xavier Lurton, *An Introduction to Underwater Acoustics* (Chichester, UK: Praxis, 2002), 115.

40. Lt. Cdr. Ryan Lilley, USN, "Recapture Wide-Area Antisubmarine Warfare," U.S. Naval Institute *Proceedings*, June 2014, https://www.usni.org/magazines/proceedings/2014/june/recapture-wide-area-antisubmarine-warfare.

41. Norman Polmar with Richard R. Burgess, *The Naval Institute Guide to the Ships and Aircraft of the U.S. Fleet*, 19th ed. (Annapolis: Naval Institute Press, 2013), 403.

42. Samuel Eliot Morison, *History of United States Naval Operations in World War II. Volume 10, The Atlantic Battle Won* (Annapolis: Naval Institute Press, 2011), 52.

43. U.S. Navy, "Vertical Launch Anti-Submarine Rocket ASROC (VLA) Missile," last modified December 6, 2013, accessed January 3, 2016, http://www.navy.mil/navydata/fact_display.asp?cid=2200&tid=1500&ct=2.

44. Adm. I. J. Galantin, USN (Ret.), *Submarine Admiral* (Chicago: University of Illinois Press, 1995), 142.

45. Lilley, "Recapture Wide-Area Antisubmarine Warfare."

46. Rear Adm. I. J. Galantin, USN, "The Future of Nuclear-Powered Submarines," U.S. Naval Institute *Proceedings*, June 1958, https://www.usni.org/magazines/proceedings/1958/june/future-nuclear-powered-submarines.

47. Rear Adm. W. J. Holland, USN (Ret.), "Submarines: Key to the Offset Strategy," U.S. Naval Institute *Proceedings*, June 2015, https://www.usni.org/magazines/proceedings/2015/june/submarines-key-offset-strategy.

48. Lilley, "Recapture Wide-Area Antisubmarine Warfare."

49. Atsushi Oi, "Why Japan's Anti-Submarine Warfare Failed," U.S. Naval Institute *Proceedings*, June 1952, https://www.usni.org/magazines/proceedings/1952/june/why-japans-anti-submarine-warfare-failed.

50. Polmar and Whitman, *Hunters and Killers: Volume II*, 212.

51. Capt. Henry J. Hendrix, USN (Ret.) (PhD), "At What Cost a Carrier?" *Center for a New American Security Disruptive Defense Papers Series*, March 2013, https://www.cnas.org/publications/reports/at-what-cost-a-carrier.

52. Roger Branfill-Cook, *Torpedo: The Complete History of the World's Most Revolutionary Naval Weapon* (Annapolis: Naval Institute Press, 2014), 229.

53. Branfill-Cook, *Torpedo*, 230.

54. Lurton, *An Introduction to Underwater Acoustics*, 115.

55. Hastings and Jenkins. *The Battle for the Falklands*, 129.

56. Lawrence Freedman, *The Official History of the Falklands Campaign. Volume II, War and Diplomacy* (London: Routledge, Taylor & Francis Group, 2005), 251.

57. Office of Program Appraisal, "Lessons of the Falklands" (Washington, DC: Department of the Navy, February 1983), 34.

58. Freedman, *The Official History of the Falklands Campaign. Volume II*, 297–98.

59. Dan van der Vat, *Stealth at Sea: The History of the Submarine* (New York: Houghton Mifflin, 1995), 340.

60. Office of Program Appraisal, "Lessons of the Falklands," 62–63.

61. Freedman, *The Official History of the Falklands Campaign. Volume II*, 429, 733.

62. Hughes, *Fleet Tactics and Coastal Combat*, 154.

63. Holland, "Submarines: Key to the Offset Strategy."

64. Freedman, *The Official History of the Falklands Campaign. Volume II*, 247.

65. Vice Adm. Michael J. Connor, USN, "Sustaining Undersea Dominance," U.S. Naval Institute *Proceedings*, June 2013, https://www.usni.org/magazines/proceedings/2013/june/sustaining-undersea-dominance.

66. Freedman, *The Official History of the Falklands Campaign. Volume II*, 298.

67. Capt. Charles H. Wilbur, USN (Ret.), "Remember the *San Luis!*" U.S. Naval Institute *Proceedings*, March 1996, 88.

68. Freedman, *The Official History of the Falklands Campaign. Volume II*, 219, 734.

69. Office of Program Appraisal, "Lessons of the Falklands," 47.

70. Keegan, *The Price of Admiralty*, 272.

71. Vandenengel, "Fighting Along a Knife Edge in the Falklands."

72. Pauline Jelinek, "AP Enterprise: Sub Attack was Near US-SKorea Drill" *Boston Globe*, June 5, 2010, http://www.boston.com/news/nation/washington /articles/2010/06/05/ap_enterprise_sub_attack_came_near_drill/.

73. The Permanent Representative of the Republic of Korea to the United Nations, "Investigation Result on the Sinking of Republic of Korea ship *Cheonan*" (New York: United Nations Security Council, June 4, 2010), 1, 5.

74. Capt. Willliam J. Toti, USN (Ret.), "The Hunt for Full-Spectrum ASW," U.S. Naval Institute *Proceedings*, June 2014, https://www.usni.org/magazines /proceedings/2014/june/hunt-full-spectrum-asw.

75. James M. McPherson, *War on the Waters: The Union and Confederate Navies, 1861–1865* (Chapel Hill: University of North Carolina Press, 2012), 5.

76. Craig L. Symonds, *The Civil War at Sea* (Santa Barbara, CA: Praeger, 2009), 153.

77. McPherson, *War on the Waters*, 210.

78. McPherson, *War on the Waters*, 213.

79. Kenneth J. Hagan, *This People's Navy: The Making of American Sea Power* (New York: Free Press, 1991), 170.

80. Capt. Wayne P. Hughes Jr., USN (Ret.), "Single-Purpose Warships for the Littorals," U.S. Naval Institute *Proceedings*, June 2014, https://www.usni.org /magazines/proceedings/2014/june/single-purpose-warships-littorals.

81. Mike Farquharson-Roberts, *A History of the Royal Navy: World War I* (London: I. B. Tauris, 2014), 77.

82. E. Michael Golda, "The Dardanelles Campaign: A Historical Analogy for Littoral Mine Warfare," *Naval War College Review*, Summer 1998: 88.

83. Farquharson-Roberts, *A History of the Royal Navy*, 79.

84. Peter Hart, *Gallipoli* (Oxford: Oxford University Press, 22011), 43.

85. Golda, "The Dardanelles Campaign," 93.

86. Golda, "The Dardanelles Campaign," 89.

87. Timothy Choi, "A Century On: The Littoral Mine Warfare Challenge," *Center for International Maritime Security*, January 27, 2016, http://cimsec. org/a-century-on-the-littoral-mine-warfare-challenge/21461.

88. Commander Timothy McGeehan, USN, and Commander Douglas Wahl, USN (Ret.), "Flash Mob in the Shipping Lane!" U.S. Naval Institute *Proceedings*, January 2016, https://www.usni.org/magazines/proceedings/2016/january /flash-mob-shipping-lane.

89. Winslow Wheeler, "More Than the Navy's Numbers Could Be Sinking," *Time*, December 4, 2012, http://nation.time.com/2012/12/04/more-than -the-navys-numbers-could-be-sinking/.

90. George W. Baer, *One Hundred Years of Sea Power: The U.S. Navy, 1890–1990* (Stanford, CA: Stanford University Press, 1993), 322.

91. Lieutenant Colonel Michael F. Trevett, USA (Ret.), "Naval Mine Warfare: Historic, Political-Military Lessons of an Asymmetric Weapon," U.S. Naval Institute *Proceedings*, August 2013, https://www.usni.org/magazines/proceedings/2013 /august/naval-mine-warfare-historic-political-military-lessons-asymmetric.

92. Scott C. Truver, "Taking Mines Seriously: Mine Warfare in China's Near Seas," *Naval War College Review*, Spring 2012: 31.

93. Golda, "The Dardanelles Campaign," 86.

94. U.S. Navy, "21st Century U.S. Navy Mine Warfare: Ensuring Global Access and Commerce" (Washington, DC: Program Executive Office, Littoral and Mine Warfare/Expeditionary Warfare Directorate, 2009), 5.

95. Truver, "Taking Mines Seriously," 31.

96. U.S. Navy, "21st Century U.S. Navy Mine Warfare," 6.

97. Lt. C. W. Nimitz, USN, "Military Value and Tactics of Modern Submarines," U.S. Naval Institute *Proceedings*, December 1912, https://www.usni.org/magazines /proceedings/1912/december-0/military-value-and-tactics-modern-submarines.

98. Padfield, *War Beneath the Sea*, 479.

99. Keegan, *The Price of Admiralty*, 274.

100. Hughes, *Fleet Tactics*, 138.

101. Cdr. G. W. Kittredge, USN, "The Impact of Nuclear Power on Submarines," U.S. Naval Institute *Proceedings*, April 1954, https://www.usni.org/magazines /proceedings/1954/april/impact-nuclear-power-submarines.

CHAPTER 7. UNDERSEA WARFARE TODAY

1. Capt. Robert C. Rubel, USN (Ret.), "The Future of the Future of Aircraft Carriers," *Naval War College Review*, Autumn 2011: 23.

2. Captain Jim Patton, USN (Ret.), "Submarine Warfare-Offense and Defense in Littorals," *Submarine Review*, December 2014, 120–21.

3. Jan Joel Andersson, "The Race to the Bottom: Submarine Proliferation and International Security," *Naval War College Review*, Winter 2015:14.

4. Capt. Henry J. Hendrix, USN, "Buy Fords, Not Ferraris," U.S. Naval Institute *Proceedings*, April 2009, https://www.usni.org/magazines/proceedings/2009 /april/buy-fords-not-ferraris.

5. Hendrix, "Buy Fords, Not Ferraris."

6. Capt. Wayne P. Hughes, USN (Ret.), *Fleet Tactics and Coastal Combat*, 2nd ed. (Annapolis: Naval Institute Press, 2000), 40, 43.

7. Hughes, *Fleet Tactics*, 44.

8. Hughes, *Fleet Tactics*, 352.

9. Xavier Lurton, *An Introduction to Underwater Acoustics* (Chichester, UK: Praxis Publishing Ltd., 2002), 112.

10. Robert J. Urick, *Principles of Underwater Sound*, 2nd ed. (New York: McGraw-Hill, 1975), 19–21.
11. Owen R. Cote Jr., "Assessing the Undersea Balance Between the U.S. and China," *SSP Working Paper*, 6.
12. Cote, "Assessing the Undersea Balance Between the U.S. and China," 5–6.
13. Lurton, *An Introduction to Underwater Acoustics*, 115.
14. Donald Ross, *Mechanics of Underwater Noise* (New York: Pergamon Press, 1976), 253.
15. Ross, *Mechanics of Underwater Noise*, 270–73.
16. Rear Adm. I. J. Galantin, USN, "The Future of Nuclear-Powered Submarines," U.S. Naval Institute *Proceedings*, June 1958, https://www.usni.org/magazines /proceedings/1958/june/future-nuclear-powered-submarines.
17. Seth Cropsey, Bryan G. McGrath, and Timothy A. Walton, "Sharpening the Spear: The Carrier, the Joint Force, and High-End Conflict," *Hudson Institute*, October 2015, 37.
18. Vice Adm. James R. Fitzgerald, USN (Ret.), "More than Submarine vs. Submarine," U.S. Naval Institute *Proceedings*, February 2013, https://www.usni.org /magazines/proceedings/2013/february/more-submarine-vs-submarine.
19. Lurton, *An Introduction to Underwater Acoustics*, 39–41.
20. Peter Howarth, *China's Rising Sea Power: The PLA Navy's Submarine Challenge* (London: Routledge, 2006), 92.
21. Urick, *Principles of Underwater Sound*, 310.
22. Owen Cote and Harvey Sapolsky, "Antisubmarine Warfare After the Cold War," *MIT Security Studies Conference Series*, October 1997, accessed March 9, 2016, https://www.files.ethz.ch/isn/92706/Antisubmarine.pdf.
23. Robert R. Mackie, "The ASW Officer: 'Jack of All Trades, Master of None,'" U.S. Naval Institute *Proceedings*, February 1972, https://www.usni.org/magazines /proceedings/1972/february/asw-officer-jack-all-trades-master-none.
24. Cdr. Robert H. Smith Jr., USN, "The Submarine's Long Shadow," U.S. Naval Institute *Proceedings*, March 1966, https://www.usni.org/magazines/proceedings/1966 /march/submarines-long-shadow.
25. Bryan Clark, "The Emerging Era in Undersea Warfare," *Center for Strategic and Budgetary Assessments*, January 22, 2015, 8.
26. Adm. James Stavridis, USN, *Destroyer Captain: Lessons of a First Command* (Annapolis: Naval Institute Press, 2008), 61.
27. Cote, "Assessing the Undersea Balance Between the U.S. and China," 5–7.
28. Lurton, *An Introduction to Underwater Acoustics*, 208.
29. Norman Polmar and Edward Whitman, *Hunters and Killers: Volume II: Anti-Submarine Warfare from 1943* (Annapolis: Naval Institute Press, 2016), 107.

30. Eric Heginbotham, et al., *The U.S.-China Military Scorecard* (Santa Monica, CA: RAND Corporation, 2015), 325.

31. Lt. Cdr. Ryan Lilley, USN, "Recapture Wide-Area Antisubmarine Warfare," U.S. Naval Institute *Proceedings*, June 2014, https://www.usni.org/magazines /proceedings/2014/june/recapture-wide-area-antisubmarine-warfare.

32. Lilley, "Recapture Wide-Area Antisubmarine Warfare."

33. Rear Adm. William J. Holland, USN (Ret.), "Fitting Submarines into the Fleet," U.S. Naval Institute *Proceedings*, June 2008, https://www.usni.org/magazines /proceedings/2008/june/fitting-submarines-fleet.

34. Holland, "Fitting Submarines into the Fleet."

35. Emphasis in original. Patton, "Submarine Warfare: Offense and Defense in Littorals," 120–21.

36. Andrew S. Erickson, Lyle J. Goldstein, William S. Murray, and Andrew R. Wilson, eds., *China's Future Nuclear Submarine Force* (Annapolis: Naval Institute Press, 2007), 71–72.

37. Erickson, Goldstein, Murray, and Wilson, eds., *China's Future Nuclear Submarine Force*, 68.

38. Capt. William Hoeft Jr., USN (Ret.), "Nobody Asked Me, But . . . In the Deep, Run Silent Again," U.S. Naval Institute *Proceedings*, January 2013, 12.

39. Norman Polmar, "U.S. Navy—Suppressing Surprise," U.S. Naval Institute *Proceedings*, October 2015, https://www.usni.org/magazines/proceedings/2015 /october/us-navy-suppressing-surprise.

40. Rear Adm. W. J. Holland, USN (Ret.), "SSN: The Queen of the Seas," *Naval War College Review*, Spring 1991: 117.

41. Norman Polmar with Richard R. Burgess, *The Naval Institute Guide to the Ships and Aircraft of the U.S. Fleet*, 19th ed. (Annapolis: Naval Institute Press, 2013), 248–50.

42. Fitzgerald, "More than Submarine vs. Submarine."

43. Patton, "Submarine Warfare: Offense and Defense in Littorals," 118.

44. Stavridis, *Destroyer Captain: Lessons of a First Command*, 61.

45. Emphasis in original. Giuseppe Fioravanzo, *A History of Naval Tactical Thought*, trans. Arthur W. Holst (Annapolis: Naval Institute Press, 1979), 209.

46. Fitzgerald, "More than Submarine vs. Submarine."

47. Hughes, *Fleet Tactics*, 40.

48. Norman Polmar, "U.S. Navy—A Paradigm Shift," U.S. Naval Institute *Proceedings*, March 2012, https://www.usni.org/magazines/proceedings/2012/march /us-navy-paradigm-shift.

49. Samuel Eliot Morison, *The Two Ocean War: A Short History of the United States Navy in the Second World War* (Boston: Little, Brown, 1963), 494.

50. Erickson, Goldstein, Murray, and Wilson, eds., *China's Future Nuclear Submarine Force*, 67.

51. Hughes, *Fleet Tactics*, 179.

52. Ian W. Toll, *Six Frigates: The Epic History of the Founding of the U.S. Navy* (New York: W. W. Norton, 2006), 341.

53. Lyle Goldstein and Shannon Knight, "Sub Force Rising," U.S. Naval Institute *Proceedings*, April 2013, https://www.usni.org/magazines/proceedings/2013/april /sub-force-rising.

54. Vice Adm. Michael J. Connor, USN, "Advancing Undersea Dominance," U.S. Naval Institute *Proceedings*, January 2015, https://www.usni.org/magazines /proceedings/2015/january/advancing-undersea-dominance.

55. Heginbotham et al., *The U.S.-China Military Scorecard*, 26.

56. Heginbotham et al., *The U.S.-China Military Scorecard*, 323.

57. Peter Dutton, Andrew S. Erickson, and Ryan Martinson, eds., *China's Near Seas Combat Capabilities* (Newport, RI: China Maritime Studies Institute, U.S. Naval War College, 2014), 17.

58. U.S. Department of Defense, "Annual Report to Congress: Military and Security Developments Involving the People's Republic of China" (Washington, DC: Office of the Secretary of Defense, 2020), 45.

59. U.S. Department of Defense, "Annual Report to Congress: Military and Security Developments Involving the People's Republic of China" (Washington, DC: Office of the Secretary of Defense, 2015), 19.

60. James C. Bussert and Bruce A. Elleman, *People's Liberation Army Navy: Combat Systems Technology, 1949–2010* (Annapolis, MD: Naval Institute Press, 2011), 62.

61. Dutton, Erickson, and Martinson, eds., *China's Near Seas Combat Capabilities*, 24.

62. Office of Naval Intelligence. "The PLA Navy: New Capabilities and Missions for the 21st Century" (Washington, DC: Office of Naval Intelligence, 2015), 19.

63. Goldstein and Knight, "Sub Force Rising."

64. Goldstein and Knight, "Sub Force Rising."

65. Erickson, Goldstein, Murray, and Wilson, eds., *China's Future Nuclear Submarine Force*, x.

66. Bussert and Elleman, *People's Liberation Army Navy*, 66.

67. Andrew S. Erickson, Lyle J. Goldstein, and Carnes Lord, eds., *China Goes to Sea: Maritime Transformation in Comparative Historical Perspective* (Annapolis: Naval Institute Press, 2009), 383.

68. Bussert and Elleman, *People's Liberation Army Navy*, 69.

69. Cote, "Assessing the Undersea Balance Between the U.S. and China," 11–12. Dutton, Erickson, and Martinson, eds., *China's Near Seas Combat Capabilities*, 19.

70. Commodore Stephen Saunders, RN, *Jane's Fighting Ships 2011–2012*, 114th ed. (Alexandria, VA: Jane's Information Group, 2011), 131.

71. Office of Naval Intelligence. "The PLA Navy: New Capabilities and Missions for the 21st Century," 19.

72. Sam LaGrone, "New Chinese Nuclear Sub Design Includes Special Operations Mini-Sub," *USNI News*, March 25, 2015, http://news.usni.org/2015/03/25 /new-chinese-nuclear-sub-design-includes-special-operations-mini-sub.

73. Office of Naval Intelligence. "The PLA Navy: New Capabilities and Missions for the 21st Century," 19.

74. U.S. Department of Defense, "Annual Report to Congress: Military and Security Developments Involving the People's Republic of China" (Washington, DC: Office of the Secretary of Defense, 2020), 45.

75. Office of Naval Intelligence. "The People's Liberation Army Navy: A Modern Navy with Chinese Characteristics" (Washington, DC: Office of Naval Intelligence, 2009), 22.

76. Bussert and Elleman, *People's Liberation Army Navy*, 67.

77. Erickson, Goldstein, Murray, and Wilson, eds., *China's Future Nuclear Submarine Force*, 60–61.

78. Saunders, *Jane's Fighting Ships 2011–2012*, 133.

79. Erickson, Goldstein, Murray, and Wilson, eds., *China's Future Nuclear Submarine Force*, 61.

80. Office of Naval Intelligence. "The PLA Navy: New Capabilities and Missions for the 21st Century," 19.

81. U.S. Department of Defense, "Annual Report to Congress: Military and Security Developments Involving the People's Republic of China" (Washington, DC: Office of the Secretary of Defense, 2015), 9.

82. Erickson, Goldstein, Murray, and Wilson, eds., *China's Future Nuclear Submarine Force*, 67.

83. Lyle Goldstein, "Emerging from the Shadows," U.S. Naval Institute *Proceedings*, April 2015, https://www.usni.org/magazines/proceedings/2015/april/emerging -shadows.

84. Dennis M. Gormley, Andrew S. Erickson, and Jingdong Yuan, "A Potent Vector: Assessing Chinese Cruise Missile Developments," *Joint Force Quarterly 75*, October 2014: 100–101.

85. Bussert and Elleman, *People's Liberation Army Navy*, 137–38.

86. Office of Naval Intelligence. "The People's Liberation Army Navy: A Modern Navy with Chinese Characteristics," 22.

87. Dutton, Erickson, and Martinson, eds., *China's Near Seas Combat Capabilities*, 19.

88. Dutton, Erickson, and Martinson, eds., *China's Near Seas Combat Capabilities,* 18.

89. Cote, "Assessing the Undersea Balance Between the U.S. and China," 3.

90. Dennis M. Gormley, Andrew S. Erickson, and Jingdong Yuan, *A Low-Visibility Force Multiplier: Assessing China's Cruise Missile Ambitions* (Washington, DC: National Defense University Press, 2014), 51.

91. Erickson, Goldstein, Murray, and Wilson, eds., *China's Future Nuclear Submarine Force,* 280–81.

92. Robert Haddick, *Fire on the Water: China, America, and the Future of the Pacific* (Annapolis: Naval Institute Press, 2014), 187.

93. U.S. Department of Defense, "Annual Report to Congress: Military and Security Developments Involving the People's Republic of China" (Washington, DC: Office of the Secretary of Defense, 2015), 35.

94. U.S. Department of Defense, "Annual Report to Congress: Military and Security Developments Involving the People's Republic of China" (Washington, DC: Office of the Secretary of Defense, 2015), 35.

95. Commander, Submarine Force, U.S. Pacific Fleet, "Attack Submarines," accessed December 19, 2020, https://www.csp.navy.mil/SUBPAC-Commands/Submarines /Attack-Submarines/.

96. Rear Adm. W. J. Holland, USN (Ret.), "Submarines: Key to the Offset Strategy," U.S. Naval Institute *Proceedings,* June 2015, https://www.usni.org/magazines /proceedings/2015/june/submarines-key-offset-strategy.

97. Cote, "Assessing the Undersea Balance Between the U.S. and China," 19.

98. Norman Polmar, *Aircraft Carriers: A History of Carrier Aviation and Its Influence on World Events, Volume II: 1946–2006* (Washington, DC: Potomac Books, 2008), 409.

99. Heginbotham et al., *The U.S.-China Military Scorecard,* 197.

100. Erickson, Goldstein, Murray, and Wilson, eds., *China's Future Nuclear Submarine Force,* 72.

101. *Game Changers: Undersea Warfare. Sea Power and Projection Forces, Before the United States House of Representatives Armed Services Committee,* 113th Cong. (October 27, 2015) (statement of Vice Adm. Michael J. Connor, USN [Ret.]).

102. Office of Naval Intelligence. "The People's Liberation Army Navy: A Modern Navy with Chinese Characteristics," 22.

103. Office of Naval Intelligence. "The Russian Navy: A Historic Transition" (Suitland, MD: Office of Naval Intelligence, December 2015), 19.

104. Saunders, *Jane's Fighting Ships 2011–2012,* 664; Prashanth Parameswaran, "Vietnam to Get Fifth Kilo Submarine from Russia in Early 2016," *Diplomat,* December 22, 2015, http://thediplomat.com/2015/12/vietnam-to-get-fifth-kilo 999-submarine-from-russia-in-early-2016/.

105. Ward Carroll and Eric Mills, hosts, "66 Years of Undersea Surveillance" with Capt. Brian Taddiken, USN, and Lt. Kristen Krock, USN, U.S. Naval Institute *Proceedings* podcast, March 2, 2021, https://www.usni.org/magazines/proceedings/the-proceedings-podcast/proceedings-podcast-210-66-years-undersea.

106. U.S. Navy, "21st Century U.S. Navy Mine Warfare: Ensuring Global Access and Commerce" (Washington, DC: Program Executive Office, Littoral and Mine Warfare/Expeditionary Warfare Directorate, 2009), 1.

107. Vice Adm. Michael J. Connor, USN, "Sustaining Undersea Dominance," U.S. Naval Institute *Proceedings*, June 2013, http://www.usni.org/magazines/proceedings/2013–06/sustaining-undersea-dominance.

108. Commander Martin Schwarz, German navy, "Future Mine Countermeasures: No Easy Solutions," *Naval War College Review*, Summer 2014: 125.

109. Andrew S. Erickson, Lyle J. Goldstein, and William S. Murray, *China's Mine Warfare: A PLA Navy 'Assassin's Mace' Capability* (Newport, RI: China Maritime Studies Institute, U.S. Naval War College, 2009), 11.

110. Office of Naval Intelligence, "The PLA Navy: New Capabilities and Missions for the 21st Century," 23–24.

111. Erickson, Goldstein, and Murray, *China's Mine Warfare*, 56.

112. Department of the Navy, "Navy Warfare Publication 3–15: Mine Warfare" (Washington, DC: Office of the Chief of Naval Operations and Headquarters U.S. Marine Corps, August 1996), 2–2.

113. Scott C. Truver, "Taking Mines Seriously: Mine Warfare in China's Near Seas," *Naval War College Review*, Spring 2012: 41.

114. Erickson, Goldstein, and Murray, *China's Mine Warfare*, 20.

115. Department of the Navy, "Navy Warfare Publication 3–15: Mine Warfare," 2–2.

116. Truver, "Taking Mines Seriously, 41; Bussert and Elleman, *People's Liberation Army Navy*, 149.

117. Erickson, Goldstein, and Murray, *China's Mine Warfare*, 14–15.

118. Truver, "Taking Mines Seriously, 34.

119. Erickson, Goldstein, and Murray, *China's Mine Warfare*, 11, 18.

120. U.S. Navy, "21st Century U.S. Navy Mine Warfare: Ensuring Global Access and Commerce," 11.

121. Erickson, Goldstein, and Murray, *China's Mine Warfare*, 26.

122. Bussert and Elleman, *People's Liberation Army Navy*, 149.

123. Erickson, Goldstein, and Murray, *China's Mine Warfare*, 28.

124. U.S. Navy, U.S. Marine Corps, and U.S. Coast Guard, "Advantage at Sea: Prevailing with Integrated All-Domain Naval Power" (Washington, DC: Office of the Deputy Chief of Naval Operations for Warfighting Development [OPNAV N7], 2020), 3.

125. Erickson, Goldstein, and Murray, *China's Mine Warfare*, 31–33.

126. Truver, "Taking Mines Seriously, 46.

127. Truver, "Taking Mines Seriously, 35.

128. Erickson, Goldstein, and Murray, *China's Mine Warfare*, 19.

129. Erickson, Goldstein, and Murray, *China's Mine Warfare*, 23.

130. Department of the Navy, "Navy Warfare Publication 3–15: Mine Warfare," 3–1.

131. Department of the Navy, "Navy Warfare Publication 3–15: Mine Warfare," 3–1.

132. Department of the Navy, "Navy Warfare Publication 3–15: Mine Warfare," 3–6.

133. Department of the Navy, "Navy Warfare Publication 3–15: Mine Warfare," 3–3.

134. Department of the Navy, "Navy Warfare Publication 3–15: Mine Warfare," 3–5.

135. Erickson, Goldstein, and Murray, *China's Mine Warfare*, 48.

136. Erickson, Goldstein, and Murray, *China's Mine Warfare*, 48.

137. Truver, "Taking Mines Seriously, 37.

138. Department of the Navy, "Navy Warfare Publication 3–15: Mine Warfare," C–6.

139. Scott C. Truver, "Wanted: U.S. Navy Mine Warfare Champion," *Naval War College Review*, Spring 2015: 120.

140. U.S. Navy, "21st Century U.S. Navy Mine Warfare: Ensuring Global Access and Commerce," 6.

141. Truver, "Wanted: U.S. Navy Mine Warfare Champion," 121.

142. Erickson, Goldstein, and Murray, *China's Mine Warfare*, 47.

143. *Game Changers: Undersea Warfare. Sea Power and Projection Forces, Before the United States House of Representatives Armed Services Committee*, 113th Cong. (October 27, 2015) (statement of Vice Adm. Michael J. Connor, USN [Ret.]).

144. *Game Changers: Undersea Warfare. Sea Power and Projection Forces, Before the United States House of Representatives Armed Services Committee*, 113th Cong. (October 27, 2015) (statement of Vice Adm. Michael J. Connor, USN [Ret.]).

145. Connor, "Sustaining Undersea Dominance."

146. Hendrix, "Buy Fords, Not Ferraris."

147. Rubel, "The Future of the Future of Aircraft Carriers," 21.

148. Cdr. Phillip E. Pournelle, USN, "When the U.S. Navy Enters the next Super Bowl, Will It Play like the Denver Broncos?" *War on the Rocks*, January 30, 2015, http://warontherocks.com/2015/01/when-the-u-s-navy-enters-the-next-super-bowl-will-it-play-like-the-denver-broncos/.

149. Lt. Jeff Vandenengel, USN, "Too Big to Sink," U.S. Naval Institute *Proceedings*, May 2017, https://www.usni.org/magazines/proceedings/2017/may/too-big-to-sink.

150. Hendrix, "Buy Fords, Not Ferraris."

151. Adm. Robert J. Natter, USN (Ret.), and Adm. Samuel J. Locklear, USN (Ret.), "Former 4-Star Fleet Commanders: Don't Give Up on Carriers," *Defense News*,

November 22, 2019, https://www.defensenews.com/naval/2019/11/22/former
-4-star-fleet-commanders-dont-give-up-on-carriers/.

PART 3. SEIZING OPPORTUNITIES

1. Rear Adm. Alfred Thayer Mahan, USN (Ret.), *Mahan on Naval Strategy* (Annapolis: Naval Institute Press, 1991), 10.

CHAPTER 8. 100,000 TONS OF INERTIA

1. Benjamin F. Armstrong, ed., *21st Century Sims* (Annapolis: Naval Institute Press, 2015), 105.
2. Lt. Cdr. Jeff Vandenengel, USN, "100,000 Tons of Inertia," U.S. Naval Institute *Proceedings*, May 2020, https://www.usni.org/magazines/proceedings/2020/may/100000-tons-inertia.
3. Upton Sinclair, *I, Candidate for Governor: And How I Got Licked* (Berkeley: University of California Press, 1934), 109.
4. Cdr. Brendan Stickles, USN, "Twilight of Manned Flight?" U.S. Naval Institute *Proceedings*, April 2016, https://www.usni.org/magazines/proceedings/2016/april/twilight-manned-flight.
5. Department of Energy and Department of the Navy, "The United States Naval Nuclear Propulsion Program" (Washington, DC: Naval Reactors, 2020), 1.
6. Ronald O'Rourke, "Navy *Ford* (CVN-78) Class Aircraft Carrier Program: Background and Issues for Congress" (Washington, DC: Congressional Research Service, December 17, 2019), 4.
7. Aircraft Carrier Industrial Base Coalition, "What We Do," accessed December 21, 2019, https://www.acibc.org/what-we-do/.
8. Mandy Smithberger, "Brass Parachutes: The Problem of the Pentagon Revolving Door," *Project on Government Oversight*, November 5, 2018.
9. John A. Tirpak, "Retired Generals Press Congress to Fund More F-35s, Discourage 'Legacy' Buy," *Air Force Magazine*, May 1, 2019, https://www.airforcemag.com/retired-generals-press-congress-to-fund-more-f-35s-discourage-legacy-buy/; Mandy Smithberger, "Officers Advocating for More F-35s Often Had Financial Stakes," *The Project on Government Oversight*, August 19, 2019, https://www.pogo.org/analysis/2019/08/officers-advocating-for-more-f-35-often-had-financial-stakes/.
10. Roxana Tiron and Tony Capaccio, "Boeing Helicopters Gain as Lawmakers Reject Army's Planned Cut," *Bloomberg Government News*, September 12, 2019, https://about.bgov.com/news/boeing-helicopters-gain-as-lawmakers-reject-armys-planned-cut/.

11. Huntington Ingalls Industries, Newport News Shipbuilding, "About Newport News Shipbuilding," accessed June 22, 2021, https://nns.huntingtoningalls.com /who-we-are/.

12. Huntington Ingalls Industries, "Huntington Ingalls Industries Chairman of the Board to Retire," press release, February 26, 2020, https://newsroom.huntingtoningalls .com/releases/chairman-fargo-retirement.

13. "Huntington Ingalls Industries Appoints Thomas Moore as New Vice President of Nuclear Operations for Nuclear and Environmental Services," Huntington Ingalls Industries Press Release, January 13, 2021, https://newsroom.huntingtoningalls.com /releases/photo-release-huntington-ingalls-industries-appoints-thomas-moore -as-new-vice-president-of-nuclear-operations-for-nuclear-and-environmental -services.

14. Huntington Ingalls Industries, "Huntington Ingalls Industries Announces DeWolfe H. Miller III as Corporate Vice President of Customer Affairs," press release, February 2, 2021, https://newsroom.huntingtoningalls.com/releases /photo-release-huntington-ingalls-industries-announces-dewolfe-h-miller-iii -as-corporate-vice-president-of-customer-affairs.

15. Huntington Ingalls Industries, "Huntington Ingalls Industries Announces James Loeblein as Corporate Vice President of Customer Affairs," press release, January 26, 2021, https://newsroom.huntingtoningalls.com/releases/photo -release-huntington-ingalls-industries-announces-james-loeblein-as-corporate -vice-president-of-customer-affairs.

16. Aircraft Carrier Industrial Base Coalition, "America's Aircraft Carriers: Action Days Infographic," September 2019, https://www.acibc.org/wp-content /uploads/2019/09/ACIBC-Actions-Days-Infographic.pdf.

17. Megan Eckstein, "After Hearings, Lawmakers Call Truman Carrier Retirement Plan 'Ridiculous,'" *USNI News*, March 28, 2019, https://news.usni.org/2019/03/28 /after-navy-hearings-lawmakers-still-call-truman-carrier-retirement-plan -ridiculous; David Larter, "Memo Reveals Pentagon Again Tried to Decommission the Carrier Truman, Cut an Air Wing, Document Shows," *Defense News*, December 30, 2019, https://www.defensenews.com/naval/2019/12/30 /the-pentagon-again-tried-to-decommission-the-carrier-truman-cut-an-air -wing-document-shows/.

18. Ben Werner, "Pence: No Early Retirement for USS Harry S. Truman," *USNI News*, April 30, 2019, https://news.usni.org/2019/04/30/pence-no-early-retirement-for -uss-harry-s-truman.

19. Title 10 U.S.C 8062(b), United States Navy: Composition; Functions. https:// www.law.cornell.edu/uscode/text/10/8062.

20. Ward Carroll and Bill Hamblet, hosts, "ADM Winnefeld Warns Winter is Coming," with Admiral James Winnefeld, USN (Ret.), U.S. Naval Institute *Proceedings* podcast, episode 171, July 12, 2020, https://www.usni.org/magazines/proceedings/the-proceedings-podcast/proceedings-podcast-episode-171-adm-winnefeld-warns.

21. Christian Brose, *The Kill Chain* (New York: Hachette Books, 2020), 221.

22. Armstrong, ed., *21st Century Sims*, 104.

23. Adm. Richard C. Macke, USN, "Now Hear This—Demise of the Aircraft Carrier? Hardly," U.S. Naval Institute *Proceedings*, October 2015, https://www.usni.org/magazines/proceedings/2015/october/now-hear-demise-aircraft-carrier-hardly.

24. "U.S. Navy Sees Chinese HGV as Part of Wider Threat," *Aviation Week and Space Technology*, January 27, 2014, http://aviationweek.com/awin/us-navy-sees-chinese-hgv-part-wider-threat. The unknown DF-21D specifications are based on those of the comparable Pershing-II missile.

25. "Building a Giant: *Gerald R. Ford (CVN 78)*," Huntington Ingalls Industries, last modified October 11, 2013, accessed December 31, 2019, http://thefordclass.com/doc/Ford-fact-sheet.pdf. Converted from stated five acres.

26. Adm. James L. Holloway III, USN (Ret.), "If the Question is China. . . .," U.S. Naval Institute *Proceedings*, January 2011, https://www.usni.org/magazines/proceedings/2011/january/if-question-china.

27. Rear Adm. Michael C. Manazir, USN. "Responsive and Relevant," U.S. Naval Institute *Proceedings*, February 2014, https://www.usni.org/magazines/proceedings/2014/february/responsive-and-relevant.

28. U.S. Navy, "Aircraft Carriers–CVN: U.S. Navy Fact File," last modified September 17, 2020, accessed January 16, 2021, https://www.navy.mil/Resources/Fact-Files/Display-FactFiles/Article/2169795/aircraft-carriers-cvn/.

29. Roger Thompson, *Lessons Not Learned: The U.S. Navy's Status Quo Culture* (Annapolis: Naval Institute Press, 2007), 65.

30. Rear Adm. I. J. Galantin, USN, "The Future of Nuclear-Powered Submarines," U.S. Naval Institute *Proceedings*, June 1958, https://www.usni.org/magazines/proceedings/1958/june/future-nuclear-powered-submarines.

31. Rebecca Grant, Adm. John Nathman, USN (Ret.), and Loren Thompson, "Get the Carriers!" U.S. Naval Institute *Proceedings*, September 2007, https://www.usni.org/magazines/proceedings/2007/september/get-carriers.

32. Rear Adm. Michael R. Groothousen, USN (Ret.), "The Enduring Relevance of America's Aircraft Carriers," *American Spectator*, March 2, 2016, http://spectator.org/65600_enduring-relevance-americas-aircraft-carriers/.

33. Adm. Robert J. Natter and Adm. Samuel J. Locklear, USN (Ret.), "Former 4-Star Fleet Commanders: Don't Give Up On Carriers," *Defense News*,

November 22, 2019, https://www.defensenews.com/naval/2019/11/22/former-4 -star-fleet-commanders-dont-give-up-on-carriers/.

34. Armstrong, ed., *21st Century Sims*, 113.

35. Rear Adm. William S. Sims, USN, "Military Conservatism," U.S. Naval Institute *Proceedings*, March 1922, https://www.usni.org/magazines/proceedings/1922 /march/military-conservatism.

36. Capt. Robert C. Rubel, USN (Ret.), "The Future of the Future of Aircraft Carriers," *Naval War College Review*, Autumn 2011: 24.

37. Carroll and Hamblet, hosts, "ADM Winnefeld Warns Winter is Coming."

38. Gen. David H. Berger, USMC, "Commandant's Planning Guidance" (Washington, DC: United States Marine Corps, July 2019), 4.

39. Paul McLeary and Lee Hudson, "How Two Dozen Retired Generals are Trying to Stop an Overhaul of the Marines," *Politico*, April 1, 2022, https://www.politico .com/news/2022/04/01/corps-detat-how-two-dozen-retired-generals-are-trying -to-stop-an-overhaul-of-the-marines-00022446.

40. Jim Webb, "Momentous Changes in the U.S. Marine Corps' Force Organization Deserve Debate," *Wall Street Journal*, March 25, 2022, https://www.wsj.com /articles/momentous-changes-in-the-marine-corps-deserve-debate-reduction -david-berger-general-11648217667?mod=opinion_lead_pos10.

41. McLeary and Hudson, "How Two Dozen Retired Generals are Trying to Stop an Overhaul of the Marines."

42. Dan Gouré, PhD, "Will Commandant Berger's Planning Guidance Mean the End of the Marine Corps?" *Real Clear Defense*, December 13, 2019, https://www .realcleardefense.com/articles/2019/12/13/will_commandant_bergers _planning_guidance_mean_the_end_of_the_marine_corps_114919.html.

43. Megan Eckstein, "Navy Focused on Strengthening Networks to Support Unmanned Operations," *USNI News*, October 27, 2020, https://news.usni.org/2020/10/27 /navy-focused-on-strengthening-networks-to-support-unmanned-operations.

44. "Statement From Deputy Secretary of Defense David L. Norquist on the Department of the Navy's Report to Congress on the Annual Long-Range Plan for Construction of Naval Vessels," U.S. Department of Defense, December 10, 2020. https://www.defense.gov/Newsroom/Releases/Release/Article /2442969/statement-from-deputy-secretary-of-defense-david-l-norquist-on-the -department-o/

45. Strategic Systems Programs, "Conventional Prompt Strike," accessed January 15, 2021, https://www.ssp.navy.mil/six_lines_of_business/cps.html.

46. Office of the Chief of Naval Operations, "Navigation Plan 2022" (Washington, DC: Department of the Navy, July 26, 2022), 8.

47. Office of the Chief of Naval Operations, "Report to Congress on the Annual Long-Range Plan for Construction of Naval Vessels" (Washington, DC: Department of the Navy, December 9, 2020), 7.

48. Office of the Chief of Naval Operations, "Navigation Plan 2022," 10.

49. Angus Ross, "Rethinking the U.S. Navy's Carrier Fleet," *War on the Rocks*, July 21, 2020, https://warontherocks.com/2020/07/rethinking-the -u-s-navys-carrier-fleet/.

50. Lt. Cdr. William S. Sims, USN, "The Inherent Tactical Qualities of All-Big-Gun, One Caliber Battleships of High Speed, Large Displacement and Gun-Power," U.S. Naval Institute *Proceedings*, October 1906, https://www.usni.org /magazines/proceedings/1906/october/inherent-tactical-qualities-all-big -gun-one-caliber-battleships.

51. Maj. Leo Spaeder, USMC, "Sir, Who Am I? An Open Letter to the Incoming Commandant of the Marine Corps," *War on the Rocks*, March 28, 2019, https:// warontherocks.com/2019/03/sir-who-am-i-an-open-letter-to-the-incoming -commandant-of-the-marine-corps/.

52. Adm. John Richardson and Lt. Ashley O'Keefe, USN, "Now Hear This—Read. Write. Fight." U.S. Naval Institute *Proceedings*, June 2016, https://www.usni.org /magazines/proceedings/2016/june/now-hear-read-write-fight.

53. Capt. Arthur H. Barber III, USN (Ret.), "Rethinking the Future Fleet," U.S. Naval Institute *Proceedings*, May 2014, http://www.usni.org/magazines/proceedings /2014–05/rethinking-future-fleet.

54. Galantin, "The Future of Nuclear-Powered Submarines."

55. Armstrong, ed., *21st Century Sims*, 105.

56. Norman Polmar, *Aircraft Carriers: A History of Carrier Aviation and Its Influence on World Events, Volume I: 1909–1945* (Washington, DC: Potomac Books, 2006), 41.

CHAPTER 9. THE SENKAKUS WAR

1. Peter Hart, *The Great War: A Combat History of the First World War* (New York: Oxford University Press, 2013), 255.

2. Michael E. O'Hanlon, *The Senkakus Paradox: Risking Great Power War Over Small Stakes* (Washington, DC: Brookings Institution Press, 2019), 3.

3. O'Hanlon, *The Senkakus Paradox*, 4–5.

4. Lieutenant Commander Jeff Vandenengel, USN, "Fighting Along a Knife Edge in the Falklands," U.S. Naval Institute *Proceedings*, December 2019, https://www.usni .org/magazines/proceedings/2019/december/fighting-along-knife-edge-falklands.

5. Office of Naval Intelligence, "The PLA Navy: New Capabilities and Missions for the 21st Century" (Washington, DC: Office of Naval Intelligence, 2015), 22.

6. U.S. Department of Defense, "Annual Report to Congress: Military and Security Developments Involving the People's Republic of China." (Washington, DC: Office of the Secretary of Defense, 2020), 51.

7. U.S. Department of Defense, "Annual Report to Congress: Military and Security Developments Involving the People's Republic of China." (Washington, DC: Office of the Secretary of Defense, 2020), 46. Office of Naval Intelligence, "The PLA Navy: New Capabilities and Missions for the 21st Century" (Washington, DC: Office of Naval Intelligence, 2015), 16.

8. Hart, *The Great War,* 255.

9. Megan Eckstein, "Navy Focused on Strengthening Networks to Support Unmanned Operations," *USNI News,* October 27, 2020, https://news.usni.org/2020/10/27/navy-focused-on-strengthening-networks-to-support-unmanned-operations.

10. As of 2022, the light amphibious warship (LAW) is an R&D program that is still under consideration, hence it does not yet have an assigned class name. In this hypothetical scenario, I took the liberty of naming it after Gen. David Berger, USMC, the commandant pushing the modernization of the Marine Corps, including the development and use of LAWs. For more background on the program, see "Report to Congress on the Light Amphibious Warship," *USNI News,* July 21, 2022, https://news.usni.org/2022/07/21/report-to-congress-on-the-light-amphibious-warship.

Bibliography

Adams, Lt. David, USN. "We Are Not Invincible." U.S. Naval Institute *Proceedings* 123, no. 5 (May 1997): 35–39.

Aircraft Carrier Industrial Base Coalition. "America's Aircraft Carriers: Action Days Infographic." September 2019. https://www.acibc.org/wp-content/uploads/2019/09/ACIBC-Actions-Days-Infographic.pdf.

———. "What We Do." Accessed December 21, 2019. https://www.acibc.orgwhat-we-do/.

Andersson, Jan Joel. "The Race to the Bottom: Submarine Proliferation and International Security." *Naval War College Review* 68, no. 1 (Winter 2015): 13–30.

Armstrong, Benjamin F., editor, *21st Century Sims*. Annapolis: Naval Institute Press, 2015.

Baer, George W. *One Hundred Years of Sea Power: The U.S. Navy, 1890–1990*. Stanford, CA: Stanford University Press, 1993.

Barber, Capt. Arthur H. III, USN (Ret.). "Redesign the Fleet." U.S. Naval Institute *Proceedings* 145, no. 1 (January 2019). https://www.usni.org/magazines/proceedings/2019/january/redesign-fleet.

———. "Rethinking the Future Fleet." U.S. Naval Institute *Proceedings* 140, no. 5 (May 2014). http://www.usni.org/magazines/proceedings/2014–05/rethinking-future-fleet.

Belote, James, and William Belote. *Typhoon of Steel: The Battle for Okinawa*. New York: Harper & Row, 1970.

Berger, Gen. David H., USMC. "Commandant's Planning Guidance." Washington, DC: United States Marine Corps, July 2019.

Binnie, Jeremy. "Sinai Militants Attack Egypt Patrol Boat." *IHS Jane's 360*, July 19, 2015. http://www.janes.com/article/53070/sinai-militants-attack-egyptian-patrol-boat.

Blair Clay Jr. *Silent Victory; The U.S. Submarine War Against Japan*. Annapolis: Naval Institute Press, 1975.

Boyd, Carl, and Akihiko Yoshida. *The Japanese Submarine Force and World War II*.

Annapolis: Naval Institute Press, 1995.

Boyer, Pelham G., and Robert S. Wood, eds. *Strategic Transformation and Naval Power in the 21st Century.* Newport, RI: Naval War College Press, 1998.

Branfill-Cook, Roger. *Torpedo: The Complete History of the World's Most Revolutionary Naval Weapon.* Annapolis: Naval Institute Press, 2014.

Broder, Jonathan. "What China's New Missiles Mean for the Future of the Aircraft Carrier." *Newsweek*, February 16, 2016. http://www.newsweek.com/china-dongfeng-21d-missile-us-aircraft-carrier-427063.

Brose, Christian. *The Kill Chain.* New York: Hachette Books, 2020.

Brown, D. K. *Nelson to Vanguard: Warship Design and Development 1923–1945.* Barnsley, England: Seaforth, 2000.

Bussert, James C., and Bruce A. Elleman. *People's Liberation Army Navy: Combat Systems Technology, 1949–2010.* Annapolis, MD: Naval Institute Press, 2011.

Carroll, Ward, and Eric Mills, hosts, "66 Years of Undersea Surveillance" with Capt. Brian Taddiken, USN, and Lt. Kristen Krock, USN. U.S. Naval Institute *Proceedings* podcast, March 2, 2021. https://www.usni.org/magazines/proceedings/the-proceedings-podcast/proceedings-podcast-210-66-years-undersea.

Carroll, Ward, and Bill Hamblet, hosts. "ADM Winnefeld Warns Winter is Coming" with Admiral James Winnefeld, USN (Ret.). U.S. Naval Institute *Proceedings* podcast, episode 171, July 12, 2020. https://www.usni.org/magazines/proceedings/the-proceedings-podcast/proceedings-podcast-episode-171-adm-winnefeld-warns.

Carter, Rear Adm. Walter E. Jr., USN. "Sea Power in the Precision-Missile Age." U.S. Naval Institute *Proceedings* 140, no. 5 (May 2014). https://www.usni.org/magazines/proceedings/2014/may/sea-power-precision-missile-age.

Cebrowski, Vice Adm. Arthur K., USN, and John H. Garstka. "Network-Centric Warfare—Its Origin and Future." U.S. Naval Institute *Proceedings* 124, no. 1 (January 1998): 28–35.

Chief of Naval Operations. "A Design for Maritime Superiority: Version 2.0." Washington, DC: Department of the Navy, December 2018.

"China Vehicle Population Hits 240 Million as Smog Engulfs Cities." *Bloomberg Business*, January 31, 2013. https://www.bloomberg.com/news/articles/2013-02-01/china-vehicle-population-hits-240-million-as-smog-engulfs-cities.

Choi, Timothy. "A Century On: The Littoral Mine Warfare Challenge." *Center for International Maritime Security*, January 27, 2016. http://cimsec.org/a-century-on-the-littoral-mine-warfare-challenge/21461.

Clark, Bryan. "The Emerging Era in Undersea Warfare." *Center for Strategic and Budgetary Assessments,* January 22, 2015.

Clark, Bryan, Peter Haynes, Bryan McGrath, Craig Hooper, Jesse Sloman, and

Timothy A. Walton. "Restoring American Seapower: A New Fleet Architecture for the United States Navy." *Center for Strategic and Budgetary Assessments*, January 2017.

Colombo, Commander Jorge Luis, ARA. "Falkland Operations I: 'Super Etendard' Naval Aircraft Operations during the Malvinas War." *Naval War College Review* 37, no. 3 (May/June 1984): 13–22.

Commander, Naval Surface Force, U.S. Pacific Fleet. "Naval Terminology." Accessed January 10, 2021. https://www.public.navy.mil/surfor/Pages/Navy-Terminology .aspx.

Commander, Submarine Force, U.S. Pacific Fleet. "Attack Submarines." Accessed December 19, 2020. https://www.csp.navy.mil/SUBPAC-Commands/Submarines /Attack-Submarines/.

Congressional Budget Office. "An Analysis of the Navy's Fiscal Year 2020 Shipbuilding Plan." Washington, DC: Congress of the United States, October 2019.

———. "The Cost of the Navy's New Frigate." Washington, DC: Congress of the United States, October 2020.

Congressional Research Service. "Conventional Prompt Global Strike and Long-Range Ballistic Missiles: Background and Issues." Washington, DC: Congressional Research Service, February 14, 2020.

———. "Navy Force Structure and Shipbuilding Plans: Background and Issues for Congress." Washington, DC: Congressional Research Service, July 28, 2020.

———. "Navy Frigate (FFG[X]) Program: Background and Issues for Congress." Washington, DC: Congressional Research Service, June 26, 2020.

Connor, Vice Adm. Michael J., USN. "Advancing Undersea Dominance." U.S. Naval Institute *Proceedings* 141, no. 1 (January 2015). https://www.usni.org/magazines /proceedings/2015/january/advancing-undersea-dominance.

Connor, Vice Adm. Michael J., USN (Ret.). *Game Changers: Undersea Warfare. Sea Power and Projection Forces, Before the United States House of Representatives Armed Services Committee*, 113th Cong., October 27, 2015.

———. "Sustaining Undersea Dominance." U.S. Naval Institute *Proceedings* 139, no. 6 (June 2013): 22–25.

Cote, Owen, and Harvey Sapolsky. "Antisubmarine Warfare After the Cold War." *MIT Security Studies Conference Series*. October 1997.

Cote, Owen R. Jr. "Assessing the Undersea Balance Between the U.S. and China." *SSP Working Paper*, February 2011.

Crooks, Lt. Cdr. Devere, USN, and Lt. Cdr. Mateo Robertaccio, USN. "The Face of Battle in the Information Age." U.S. Naval Institute *Proceedings* 141, no. 7 (July 2015). http://www.usni.org/magazines/proceedings/2015–07/face-battle-information-age.

Cropsey, Seth, Bryan G. McGrath, and Timothy A. Walton. "Sharpening the Spear: The Carrier, the Joint Force, and High-End Conflict." *Hudson Institute*, October 2015.

Daughan, George C. *1812: The Navy's War*. New York: Basic Books, 2011.

Davis, Brian L. *Qaddafi, Terrorism, and the Origins of the U.S. Attack on Libya*. New York: Praeger, 1990.

Defense Intelligence Agency. "Russia Military Power: Building a Military to Support Great Power Aspirations." Washington, DC: Defense Intelligence Agency, 2017.

U.S. Department of Defense, "Statement from Deputy Secretary of Defense David L. Norquist on the Department of the Navy's Report to Congress on the Annual Long-Range Plan for Construction of Naval Vessels," December 10, 2020. https://www.defense.gov/Newsroom/Releases/Release/Article/2442969/statement-from-deputy-secretary-of-defense-david-l-norquist-on-the-department-o/.

Department of Energy and Department of the Navy. "The United States Naval Nuclear Propulsion Program." Washington, DC: Naval Reactors, 2020.

Department of the Navy. "2017 U.S. Navy Program Guide." Washington, DC: Department of the Navy, 2017.

———. "Navy Warfare Publication 3–15: Mine Warfare." Washington, DC: Office of the Chief of Naval Operations and Headquarters U.S. Marine Corps, August 1996.

Director, Operational Test and Evaluation. "FY14 Annual Report." Washington, DC: Office of the Director, Operational Test and Evaluation, January 2015.

Dutton, Peter, Andrew S. Erickson, and Ryan Martinson, eds. *China's Near Seas Combat Capabilities*. Newport, RI: China Maritime Studies Institute, U.S. Naval War College, 2014.

Eckstein, Megan. "After Hearings, Lawmakers Call Truman Carrier Retirement Plan 'Ridiculous,'" *USNI News*, March 28, 2019. https://news.usni.org/2019/03/28/after-navy-hearings-lawmakers-still-call-truman-carrier-retirement-plan-ridiculous.

———. "Navy Focused on Strengthening Networks to Support Unmanned Operations." *USNI News*, October 27, 2020. https://news.usni.org/2020/10/27/navy-focused-on-strengthening-networks-to-support-unmanned-operations.

———. "Pentagon Leaders Have Taken Lead in Crafting Future Fleet from Navy." *USNI News*, June 24, 2020. https://news.usni.org/2020/06/24/pentagon-leaders-have-taken-lead-in-crafting-future-fleet-from-navy.

Edwards, Joshua J., and Captain Dennis M. Gallagher, USN. "Mine and Undersea Warfare for the Future." U.S. Naval Institute *Proceedings* 140, no. 8 (August 2014): https://www.usni.org/magazines/proceedings/2014/august/mine-and-undersea-warfare-future.

Erickson, Andrew S. *Chinese Anti-Ship Ballistic Missile Development: Drivers, Trajectories and Strategic Implications.* Washington, DC: Jamestown Foundation, 2013.

Erickson, Andrew S., Lyle J. Goldstein, and Carnes Lord, eds. *China Goes to Sea: Maritime Transformation in Comparative Historical Perspective.* Annapolis: Naval Institute Press, 2009.

Erickson, Andrew S., Lyle J. Goldstein, and William S. Murray. *China's Mine Warfare: A PLA Navy 'Assassin's Mace' Capability.* Newport, RI: China Maritime Studies Institute, U.S. Naval War College, 2009.

Erickson, Andrew S., Lyle J. Goldstein, William S. Murray, and Andrew R. Wilson, eds. *China's Future Nuclear Submarine Force.* Annapolis: Naval Institute Press, 2007.

Erickson, Andrew S., and Conor M. Kennedy. "China's Maritime Militia." Center for Naval Analyses, 7 March 2016. www.cna.org/cna_files/pdf/Chinas-Maritime -Militia.pdf.

Fabey, Michael. "US Navy CNO Cites Benefits of 'Aviation Combatants' for Future Force." *Janes*, October 14, 2020. https://www.janes.com/defence-news/news-detail /us-navy-cno-cites-benefits-of-aviation-combatants-for-future-force.

Fanell, Captain James, USN. "Chinese Navy: Operational Challenge or Potential Partner?" Panel at Naval Institute-AFCEA West 2014 Conference, San Diego, California, January 31, 2013.

Farquharson-Roberts, Mike. *A History of the Royal Navy: World War I.* London: I. B. Tauris, 2014.

Felger, Daniel G. *Engineering for Officer of the Deck.* Annapolis: Naval Institute Press, 1979.

Fioravanzo, Giuseppe. Translated by Arthur W. Holst. *A History of Naval Tactical Thought.* Annapolis: Naval Institute Press, 1979.

Fitzgerald, Vice Adm. James R., USN (Ret.). "More than Submarine vs. Submarine." U.S. Naval Institute *Proceedings* 139, no. 2 (February 2013): 32–37.

Flintham, Victor. *Air Wars and Aircraft: A Detailed Record of Air Combat, 1945 to the Present.* New York: Facts on File, 1990.

Ford, Worthington Chauncey. *The Writings of George Washington.* New York: Knickerbocker Press, 1891.

Freedberg, Sydney J. Jr. "Polmar's Navy: Trade LCS & Carriers for Frigates and Amphibs." *Breaking Defense*, December 18, 2015. http://breakingdefense .com/2015/12/polmars-navy-trade-lcs-carriers-for-frigates-amphibs/.

Freedman, Lawrence. *The Official History of the Falklands Campaign. Volume II, War and Diplomacy.* London: Routledge, Taylor & Francis Group, 2005.

Friedman, Norman. "F-35B Expands Definition of 'Aircraft Carriers.'" U.S. Naval Institute *Proceedings* 145, no. 2 (February 2019). https://www.usni.org/magazines /proceedings/2019/february/f-35b-expands-definition-aircraft-carriers.

Galantin, Rear Adm. I. J., USN. "The Future of Nuclear-Powered Submarines." U.S. Naval Institute *Proceedings* 84, no. 6 (June 1958): 23–35.

Galantin, Adm. I. J., USN (Ret.). *Submarine Admiral: From Battlewagons to Ballistic Missiles*. Chicago: University of Illinois Press, 1995.

General Dynamics, Electric Boat. "Our Submarines." Accessed January 2, 2016. http://www.gdeb.com/about/oursubmarines/.

Gibran, Daniel K. *The Falklands War: Britain Versus the Past in the South Atlantic*. London: McFarland, 1998.

Golda, E. Michael. "The Dardanelles Campaign: A Historical Analogy for Littoral Mine Warfare." *Naval War College Review* 51, no. 3 (Summer 1998): 82–96.

Goldstein, Lyle. "Emerging from the Shadows." U.S. Naval Institute *Proceedings* 141, no. 4 (April 2015): https://www.usni.org/magazines/proceedings/2015/april/emerging-shadows.

Goldstein, Lyle, and Shannon Knight. "Sub Force Rising." U.S. Naval Institute *Proceedings* 139, no. 4 (April 2013): 40–44.

Gormley, Dennis M., Andrew S. Erickson, and Jingdong Yuan. *A Low-Visibility Force Multiplier: Assessing China's Cruise Missile Ambitions*. Washington, DC: National Defense University Press, 2014.

Gormley, Dennis M., Andrew S. Erickson, and Jingdong Yuan. "A Potent Vector: Assessing Chinese Cruise Missile Developments." *Joint Force Quarterly* 75, no. 4 (October 2014): 98–105.

Gouré, Dan, PhD. "Will Commandant Berger's Planning Guidance Mean the End of the Marine Corps?" *Real Clear Defense*, December 13, 2019. https://www.realcleardefense.com/articles/2019/12/13/will_commandant_bergers_planning_guidance_mean_the_end_of_the_marine_corps_114919.html.

Grant, Rebecca, Adm. John Nathman, USN (Ret.), and Loren Thompson. "Get the Carriers!" U.S. Naval Institute *Proceedings* 133, no. 9 (September 2007): 38–42.

Greenert, Adm. Jonathan, USN. "Chief of Naval Operations Remarks." Slade Gorton International Policy Center Luncheon. Seattle, WA, September 24, 2013.

———. "Navy 2025: Forward Warfighters." U.S. Naval Institute *Proceedings* 137, no. 12 (December 2011): 18–23.

———. "Payloads Over Platforms: Charting a New Course." U.S. Naval Institute *Proceedings* 138, no. 7 (July 2012): 16–23. https://www.usni.org/magazines/proceedings/2012/july/payloads-over-platforms-charting-new-course.

Gronlund, Lisbeth, David C. Wright, George N. Lewis, and Philip E. Coyle III. "Technical Realities: An Analysis of the 2004 Deployment of a U.S. National Missile Defense System." Cambridge, MA: Union of Concerned Scientists, May 2004.

Groothousen, Rear Adm. Michael R., USN (Ret.). "The Enduring Relevance of America's Aircraft Carriers." *American Spectator*, March 2, 2016. http://spectator.org/65600_enduring-relevance-americas-aircraft-carriers/.

Haddick, Robert. *Fire on the Water: China, America, and the Future of the Pacific.* Annapolis: Naval Institute Press, 2014.

Hagan, Kenneth J. *This People's Navy: The Making of American Sea Power.* New York: Free Press, 1991.

Hagt, Eric, and Matthew Durnin. "China's Antiship Ballistic Missile: Developments and Missing Links." *Naval War College Review* 62, no. 4 (Autumn 2009): 87–113.

Hara, Captain Tameichi, IJN, with Fred Saito and Roger Pineau. *Japanese Destroyer Captain: Pearl Harbor, Guadalcanal, Midway—as Seen Through Japanese Eyes.* Annapolis: Naval Institute Press, 1967.

Harel, Amos, and Avi Issacharoff. *34 Days: Israel, Hezbollah, and the War in Lebanon.* New York: Palgreave MacMillan, 2008.

Harris, Admiral Harry B. Jr. USN. *U.S. Pacific Command Posture Testimony, Before the United States House of Representatives Armed Services Committee,* 115th Cong., April 26, 2017.

Harris, Capt. R. Robinson, USN (Ret.), Andrew Kerr, Kenneth Adams, Christopher Abt, Michael Venn, and Colonel T.X. Hammes, USMC (Ret.). "Converting Merchant Ships to Missile Ships for the Win." U.S. Naval Institute *Proceedings* 145, no. 1 (January 2019): https://www.usni.org/magazines/proceedings/2019/january/converting-merchant-ships-missile-ships-win.

Hart, Peter. *Gallipoli.* Oxford: Oxford University Press, 2011.

———. *The Great War: A Combat History of the First World War.* New York: Oxford University Press, 2013.

Harvey, Adm. John, Jr. USN (Ret.), Capt. Wayne Hughes Jr., USN (Ret.), Capt. Jeffrey Kline, USN (Ret.), and Lt. Zachary Schwartz, USN. "Sustaining American Maritime Influence." U.S. Naval Institute *Proceedings* 139, no. 9 (September 2013): 46–51.

Hastings, Max, and Simon Jenkins. *The Battle for the Falklands.* New York: W. W. Norton, 1983.

Heginbotham, Eric, Michael Nixon, Forrest E. Morgan, Jacob L. Heim, Jeff Hagen, Sheng Tao Li, Jeffrey Engstrom, Martin C. Libicki, Paul DeLuca, David Shlapak, David R. Frelinger, Burgess Laird, Kyle Brady, and Lyle J. Morris. *The U.S.-China Military Scorecard.* Santa Monica, CA: RAND Corporation, 2015.

Henderson, Capt. Arnold, USN (Ret.). "Training Against the Navy's #1 Threat." U.S. Naval Institute *Proceedings* 123, no. 9 (September 1997): 104–107.

Hendrix, Capt. Henry J. "Buy Fords, Not Ferraris." U.S. Naval Institute *Proceedings* 135, no. 4 (April 2009): 52–57.

Hendrix, Capt. Henry J., USN (PhD). "At What Cost a Carrier?" *Center for a New American Security Disruptive Defense Papers Series*, March 2013.

Hendrix, Dr. Jerry. "Retreat From Range: The Rise and Fall of Carrier Aviation." *Center for a New American Security,* October 2015.

Hendrix, Dr. Jerry, and Cdr. Benjamin Armstrong, USN. "The Presence Problem: Naval Presence and National Strategy." *Center for a New American Security*, January 2016.

Hesser Captain William A. Jr., USN. "Command Investigation into the Target Drone Malfunction and Strike of the USS Chancellorsville (CG 62) on 16 November 2013." Pearl Harbor, HI: Commander, U.S. Pacific Fleet, December 20, 2013.

Hind, Andrew. "The Cruise Missile Comes of Age." *Naval History Magazine* 22, no. 5 (October 2008): https://www.usni.org/magazines/naval-history-magazine/2008/october/cruise-missile-comes-age.

Hoeft, Capt. William Jr., USN (Ret.). "Nobody Asked Me, But . . . In the Deep, Run Silent Again." U.S. Naval Institute *Proceedings* 139, no. 1 (January 2013): 12.

Holland, Rear Adm. W. J., USN (Ret.). "SSN: The Queen of the Seas." *Naval War College Review* 44, no. 2 (Spring 1991): 113–18.

———. "Submarines: Key to the Offset Strategy." U.S. Naval Institute *Proceedings* 141, no. 6 (June 2015). https://www.usni.org/magazines/proceedings/2015/june/submarines-key-offset-strategy.

———. "Fitting Submarines into the Fleet." U.S. Naval Institute *Proceedings* 134, no. 6 (June 2008): 32–36.

Holloway III, Adm. James L., USN (Ret.). *Aircraft Carriers at War*. Annapolis: Naval Institute Press, 2007.

———. "If the Question is China…," U.S. Naval Institute *Proceedings* 137, no. 1 (January 2011): 54–57.

Hone, Trent. *Learning War*. Annapolis: Naval Institute Press, 2018.

Howarth, Peter. *China's Rising Sea Power: The PLA Navy's Submarine Challenge*. London: Routledge, 2006.

Hoyler, Marshall. "China's 'Antiaccess' Ballistic Missiles and U.S. Active Defense." *Naval War College Review* 63, no. 4 (Autumn 2010): 84–105.

Hughes, Capt. Wayne P. Jr., USN (Ret.). *Fleet Tactics and Coastal Combat*, 2nd ed. Annapolis: Naval Institute Press, 2000.

———. "The New Navy Fighting Machine: A Study of the Connections Between Contemporary Policy, Strategy, Sea Power, Naval Operations, and the Composition of the United States Fleet." Naval Postgraduate School, August 2009.

———. "Restore a Distributable Naval Air Force." U.S. Naval Institute *Proceedings* 145, no. 4 (April 2019): 24–27.

———. "Single-Purpose Warships for the Littorals." U.S. Naval Institute *Proceedings* 140, no. 6 (June 2014): https://www.usni.org/magazines/proceedings/2014/june/single-purpose-warships-littorals.

———. "22 Questions for Streetfighter." U.S. Naval Institute *Pr-ceedings* 126, no. 2 (February 2000): 46–49. https://www.usni.org/magazines/proceedings/2000/february/22-questions-streetfighter.

Huntington Ingalls Industries. "Building a Giant: *Gerald R. Ford (CVN 78)*." Last modified October 11, 2013, accessed August 7, 2015. http://thefordclass.com/doc /Ford-fact-sheet.pdf.

———. "Huntington Ingalls Industries Announces DeWolfe H. Miller III as Corporate Vice President of Customer Affairs." February 2, 2021. https://newsroom .huntingtoningalls.com/releases/photo-release-huntington-ingalls-industries -announces-dewolfe-h-miller-iii-as-corporate-vice-president-of-customer-affairs.

———. "Huntington Ingalls Industries Announces James Loeblein as Corporate Vice President of Customer Affairs." January 26, 2021. https://newsroom.huntingtoningalls .com/releases/photo-release-huntington-ingalls-industries-announces-james -loeblein-as-corporate-vice-president-of-customer-affairs.

———. "Huntington Ingalls Industries Appoints Thomas Moore as New Vice President of Nuclear Operations for Nuclear and Environmental Services." January 13, 2021. https://newsroom.huntingtoningalls.com/releases/photo-release -huntington-ingalls-industries-appoints-thomas-moore-as-new-vice-president -of-nuclear-operations-for-nuclear-and-environmental-services.

———. "Huntington Ingalls Industries Chairman of the Board to Retire." Press release. February 26, 2020. https://newsroom.huntingtoningalls.com/releases /chairman-fargo-retirement.

Huntington Ingalls Industries, Newport News Shipbuilding. "About Newport News Shipbuilding." Accessed June 22, 2021. https://nns.huntingtoningalls.com/ who-we-are/.

Jelinek, Pauline. "AP Enterprise: Sub Attack was Near US–SKorea Drill." *Boston Globe*, June 5, 2010. http://www.boston.com/news/nation/washington/articles/2010/06/05 /ap_enterprise_sub_attack_came_near_drill/.

Johnson, Stuart E., and Vice Adm. Arthur K. Cebrowski, USN (Ret.). "Alternative Fleet Architecture Design." Center for Technology and National Security Policy, National Defense University, August 2005.

Keegan, John. *The Price of Admiralty: The Evolution of Naval Warfare*. New York: Viking Penguin, 1989.

Kittredge, Commander G. W., USN. "The Impact of Nuclear Power on Submarines." U.S. Naval Institute *Proceedings* 80, no. 4 (April 1954): 419–25.

LaGrone, Sam. "New Chinese Nuclear Sub Design Includes Special Operations Mini-Sub." *USNI News*, March 25, 2015. http://news.usni.org/2015/03/25 /new-chinese-nuclear-sub-design-includes-special-operations-mini-sub.

———. "USS *Mason* Fired 3 Missiles to Defend From Yemen Cruise Missile Attacks." *USNI News*, October 11, 2016. https://news.usni.org/2016/10/11/uss-mason -fired-3-missiles-to-defend-from-yemen-cruise-missiles-attack.

————. "Warship *Moskva* Was Blind to Ukrainian Missile Attack, Analysis Shows." *USNI News*, May 5, 2022. https://news.usni.org/2022/05/05/warship -moskva-was-blind-to-ukrainian-missile-attack-analysis-shows.

Lambert, Nicholas A. "What Is a Navy For?" U.S. Naval Institute *Proceedings* 147, no. 4, (April 2021). https://www.usni.org/magazines/proceedings/2021/april/what-navy.

Larter, David. "Memo Reveals Pentagon Again Tried to Decommission the Carrier Truman, Cut an Air Wing." *Defense News*, December 30, 2019. https://www. defensenews.com/naval/2019/12/30/the-pentagon-again-tried-to-decommission -the-carrier-truman-cut-an-air-wing-document-shows/.

————. "U.S. Navy Looks to Hire Thousands More Sailors as Service Finds Itself 9,000 Sailors Short at Sea." *Navy Times*, February 10, 2020. https://www.navytimes.com /smr/federal-budget/2020/02/11/us-navy-looks-to-hire-thousands-more-sailors- as-service-finds-itself-9000-sailors-short-at-sea/.

Lecki, Robert. *Okinawa: The Last Battle of World War II.* New York: Penguin Group, 1995.

Levinson, Jeffrey L., and Randy L. Edwards. *Missile Inbound: The Attack on the Stark in the Persian Gulf.* Annapolis: Naval Institute Press, 1997.

Lewis, George N., and Theodore A. Postol. "A Flawed and Dangerous U.S. Missile Defense Plan." *Arms Control Today.* May 5, 2010. https://www.armscontrol.org /act/2010_05/Lewis-Postol.

Lilley, Lt. Cdr. Ryan, USN. "Recapture Wide-Area Antisubmarine Warfare." U.S. Naval Institute *Proceedings* 140, no. 6 (June 2014). https://www.usni.org/magazines /proceedings/2014/june/recapture-wide-area-antisubmarine-warfare.

Lockheed Martin. "F-35 Weaponry." Accessed August 21, 2020. https://www.f35.com /about/carrytheload/weaponry#:~:text=In%20stealth%20mode%2C%20the%20 F,battle%20to%20finish%20the%20fight.

Lockheed Martin. "F-35C Carrier Variant." Accessed May 24, 2015. http://www .lockheedmartin.com/us/products/f35/f-35c-carrier-variant.html.

Lurton, Xavier. *An Introduction to Underwater Acoustics.* Chichester, UK: Praxis, 2002.

Macke, Adm. Richard C., USN. "Now Hear This—Demise of the Aircraft Carrier? Hardly." U.S. Naval Institute *Proceedings* 141, no. 10 (October 2015). https://www.usni.org/magazines/proceedings/2015/october/now-hear-demise -aircraft-carrier-hardly.

Mackie, Robert R. "The ASW Officer: 'Jack of All Trades, Master of None.'" U.S. Naval Institute *Proceedings* 98, no. 2 (February 1972): 34–40.

Mahan, Alfred Thayer. *Armaments and Arbitration: The Place of Force in the International Relations of States.* New York: Harper and Brothers, 1912.

————. *Lessons of the War with Spain.* London: Sampson Low, Marston, 1899.

———. *Mahan on Naval Strategy*. Annapolis: Naval Institute Press, 1991.

Manazir, Rear Admiral Michael C., USN. "Responsive and Relevant." U.S. Naval Institute *Proceedings* 140, no. 2 (February 2014): https://www.usni.org/magazines/proceedings/2014/february/responsive-and-relevant.

Marolda, Edward J., and Robert J. Schneller Jr. *Shield and Sword: The United States Navy and the Persian Gulf War*. Washington, DC: Naval Historical Center, Department of the Navy, 1998.

Martin, Bradley, and Michael E. McMahon. *Future Aircraft Carrier Options*. Santa Monica, CA: RAND, 2017.

Mazarr, Michael J. *Understanding Deterrence*. Santa Monica, CA: RAND, 2018.

McGeehan, Cdr. Timothy, USN, and Cdr. Douglas Wahl, USN (Ret.). "Flash Mob in the Shipping Lane!" U.S. Naval Institute *Proceedings* 142, no. 1 (January 2016): https://www.usni.org/magazines/proceedings/2016/january/flash-mob-shipping-lane.

McLeary, Paul, and Lee Hudson. "How Two Dozen Retired Generals are Trying to Stop an Overhaul of the Marines." *Politico*, April 1, 2022. https://www.politico.com/news/2022/04/01/corps-detat-how-two-dozen-retired-generals-are-trying-to-stop-an-overhaul-of-the-marines-00022446.

McPherson, James M. *War on the Waters: The Union and Confederate Navies, 1861–1865*. Chapel Hill, NC: University of North Carolina Press, 2012.

Morgan, James. *Theodore Roosevelt: The Boy and the Man*. New York: MacMillan, 1907.

Morison, Samuel Eliot. *History of United States Naval Operations in World War II*. Volume 6, *Breaking the Bismarcks Barrier*. Annapolis: Naval Institute Press, 1950.

———. *History of United States Naval Operations in World War II*. Volume 10, *The Atlantic Battle Won*. Annapolis: Naval Institute Press, 2011.

———. *The Two Ocean War: A Short History of the United States Navy in the Second World War*. Boston: Little, Brown, 1963.

Nakashima, Ellen, and Craig Timberg. "Russian government hackers are behind a broad espionage campaign that has compromised U.S. agencies, including Treasury and Commerce." *Washington Post*, December 14, 2020. https://www.washingtonpost.com/national-security/russian-government-spies-are-behind-a-broad-hacking-campaign-that-has-breached-us-agencies-and-a-top-cyber-firm/2020/12/13/d5a53b88-3d7d-11eb-9453-fc36ba051781_story.html.

National Research Council. *Responding to Capability Surprise: A Strategy for U.S. Forces*. Washington, DC: National Academies Press, 2013.

Natter, Adm. Robert J., USN (Ret.), and Adm. Samuel J. Locklear, USN (Ret.). "Former 4-Star Fleet Commanders: Don't Give Up on Carriers." *Defense News*, November 22, 2019. https://www.defensenews.com/naval/2019/11/22/former-4-star-fleet-commanders-dont-give-up-on-carriers/.

Naval Air Systems Command. "MQ-8 Fire Scout." Accessed July 12, 2016. http://www.navair.navy.mil/index.cfm?fuseaction=home.display&key=8250AFBA-DF2B-4999-9EF3-0B0E46144D03.

Naval History and Heritage Command. "Sterett III (DLG-31)." Last modified April 28, 2020, accessed July 5, 2021. https://www.history.navy.mil/research/histories/ship-histories/danfs/s/sterett-iii.html.

———. "United States Submarine Losses World War II: Introduction." Last modified July 23, 2015, accessed November 28, 2015. http://www.history.navy.mil/research/library/online-reading-room/title-list-alphabetically/u/united-states-submarine-losses/introduction.html.

———. "U.S. Ship Force Levels." Last modified November 17, 2017, accessed July 9, 2020. https://www.history.navy.mil/research/histories/ship-histories/us-ship-force-levels.html.

Niestlé, Axel. *German U-Boat Losses During World War II: Details of Destruction.* London: Frontline Books, 2014.

Nimitz, Lieutenant C. W., USN. "Military Value and Tactics of Modern Submarines." U.S. Naval Institute *Proceedings* 38, no. 4 (December 1912): 1194–1211.

Norton, Steven, and Clint Boulton. "Years of Tech Mismanagement Led to OPM Breach, Resignation of Chief." *Wall Street Journal,* July 10, 2015. http://blogs.wsj.com/cio/2015/07/10/years-of-tech-mismanagement-led-to-opm-breach-resignation-of-chief/.

O'Hanlon, Michael E. *The Senkakus Paradox: Risking Great Power War Over Small Stakes.* Washington, DC: Brookings Institution Press, 2019.

O'Rourke, Ronald. "Navy Aegis Ballistic Missile Defense (BMD) Program: Background and Issues for Congress." Washington, DC: Congressional Research Service, March 4, 2015.

———. "Navy Aegis Ballistic Missile Defense (BMD) Program: Background and Issues for Congress." Washington, DC: Congressional Research Service, December 23, 2020.

———. "Navy Ford (CVN-78) Class Aircraft Carrier Program: Background and Issues for Congress." Washington, DC: Congressional Research Service, December 17, 2019.

———. "Navy Ford (CVN-78) Class Aircraft Carrier Program: Background and Issues for Congress." Washington, DC: Congressional Research Service, June 8, 2020.

———. "The Tanker War." U.S. Naval Institute *Proceedings* 114, no. 5 (May 1988): 29–34.

Office of the Chief of Naval Operations. "Navigation Plan 2022." Washington, DC: Department of the Navy, July 26, 2022.

———. "Report to Congress on the Annual Long-Range Plan for Construction of Naval Vessels." Washington, DC: Department of the Navy, December 9, 2020.

———. "Report to Congress on the Annual Long-Range Plan for Construction of Naval Vessels for Fiscal Year 2020." Washington, DC: Department of the Navy, March 2019.

Office of Naval Intelligence. "The People's Liberation Army Navy: A Modern Navy with Chinese Characteristics." Washington, DC: Office of Naval Intelligence, 2009.

Office of Naval Intelligence. "The PLA Navy: New Capabilities and Missions for the 21st Century." Washington, DC: Office of Naval Intelligence, 2015.

———. "The Russian Navy: A Historic Transition." Suitland, MD: Office of Naval Intelligence, December 2015.

Office of Program Appraisal. "Lessons of the Falklands." Washington, DC: Department of the Navy, February 1983.

Oi, Atsushi. "Why Japan's Anti-Submarine Warfare Failed." U.S. Naval Institute *Proceedings* 78, no. 6 (June 1952): 587–601.

Padfield, Peter. *War Beneath the Sea: Submarine Conflict During World War II*. New York: John Wiley & Son, 1995.

Paine, Lincoln. *The Sea and Civilization: A Maritime History of the World*. New York: Vintage Books, 2013.

Palit, Major General D. K., Vr C (Ret.). *The Lightning Campaign: The Indo-Pakistan War 1971*. New Delhi: Thomson Press (India), 1972.

Parameswaran, Prashanth. "Vietnam to Get Fifth *Kilo* Submarine from Russia in Early 2016." *Diplomat*, December 22, 2015. http://thediplomat.com/2015/12/vietnam-to-get-fifth-kilo-submarine-from-russia-in-early-2016/.

Parrish, Thomas. *The Submarine: A History*. New York: Penguin, 2004.

Patch, Cdr. John, USN (Ret.). "Fortress at Sea? The Carrier Invulnerability Myth." U.S. Naval Institute *Proceedings* 136, no. 1 (January 2010): 16–20.

"Patriotisms." *Science*. April 17, 1992. http://www.sciencemag.org/content/256/5055/312.full.pdf?sid=8515261b-059b-4e73-93fd-0ea4b44daab4.

Patton, Capt. Jim, USN (Ret.). "Submarine Warfare-Offense and Defense in Littorals." *Submarine Review* (December 2014): 115–22.

Polmar, Norman. *Aircraft Carriers: A History of Carrier Aviation and Its Influence on World Events, Volume I: 1909–1945* Washington, DC: Potomac Books, 2006.

———. *Aircraft Carriers: A History of Carrier Aviation and Its Influence on World Events, Volume II: 1946–2006*. Washington, DC: Potomac Books, 2008.

———. "U.S. Navy—A Paradigm Shift." U.S. Naval Institute *Proceedings* 138, no. 3 (March 2012): 88–89.

Polmar, Norman. "U.S. Navy—Suppressing Surprise." U.S. Naval Institute *Proceedings* 141, no. 10 (October 2015): https://www.usni.org/magazines/proceedings/2015/october/us-navy-suppressing-surprise.

Polmar, Norman, and Edward Whitman. *Hunters and Killers: Volume 1: Anti-Submarine Warfare from 1776 to 1943.* Annapolis: Naval Institute Press, 2015.

———. *Hunters and Killers: Volume 2: Anti-Submarine Warfare from 1943.* Annapolis: Naval Institute Press, 2016.

Polmar, Norman, with Richard R. Burgess. *The Naval Institute Guide to the Ships and Aircraft of the U.S. Fleet,* 19th ed. Annapolis: Naval Institute Press, 2013.

Pournelle, Cdr. Phillip E., USN. "The Deadly Future of Littoral Sea Control." U.S. Naval Institute *Proceedings* 141, no. 7 (July 2015). https://www.usni.org/magazines/proceedings/2015/july/deadly-future-littoral-sea-control.

———. "The Rise of the Missile Carriers." U.S. Naval Institute *Proceedings* 139, no. 5 (May 2013): 30–34.

———. "We Need a Balanced Fleet for Naval Supremacy." *Information Dissemination,* November 4, 2013. http://www.informationdissemination.net/2013/11/we-need-balanced-fleet-for-naval.html.

———. "When the U.S. Navy Enters the next Super Bowl, Will It Play like the Denver Broncos?" *War on the Rocks,* January 30, 2015, http://warontherocks.com/2015/01/when-the-u-s-navy-enters-the-next-super-bowl-will-it-play-like-the-denver-broncos/.

Prados, John. *Vietnam: The History of an Unwinnable War, 1945–1975.* Lawrence: University Press of Kansas, 2009.

Pradun, Vitaliy O. "From Bottle Rockets to Lightning Bolts: China's Missile Revolution and Strategy Against U.S. Military Intervention." *Naval War College Review* 64, no. 2 (Spring 2011): 7–33.

Rabinovich, Abraham. "From 'Futuristic Whimsy' to Naval Reality," *Naval History Magazine* 28, no. 3 (June 2014). https://www.usni.org/magazines/naval-history-magazine/2014/june/futuristic-whimsy-naval-reality.

Rajaee, Farhang. *The Iran Iraq War: The Politics of Aggression.* Gainesville: University Press of Florida, 1993.

RAND Corporation. "The New Face of Naval Strike Warfare." *RAND National Defense Research Institute Research Brief,* 2005.

"Report to Congress on the Light Amphibious Warship." *USNI News,* July 21, 2022. https://news.usni.org/2022/07/21/report-to-congress-on-the-light-amphibious-warship.

Richardson, Adm. John, and Lt. Ashley O'Keefe, USN. "Now Hear This—Read. Write. Fight." U.S. Naval Institute *Proceedings* 142, no. 6 (June 2016). https://www.usni.org/magazines/proceedings/2016/june/now-hear-read-write-fight.

Risen, James. "U.S. Warns China on Taiwan, Sends Warships to Area." *Los Angeles Times,* March 11, 1996. https://www.latimes.com/archives/la-xpm-1996-03-11-mn-45722-story.html.

Ross, Angus. "Rethinking the U.S. Navy's Carrier Fleet." *War on the Rocks*, July 21, 2020. https://warontherocks.com/2020/07/rethinking-the-u-s-navys-carrier-fleet/.

Ross, Donald. *Mechanics of Underwater Noise*. New York: Pergamon Press, 1976.

Rowden, Vice Adm. Thomas, Rear Adm. Peter Gumataotao, and Rear Adm. Peter Fanta, USN. "Distributed Lethality." U.S. Naval Institute *Proceedings* 141, no. 1 (January 2015). https://www.usni.org/magazines/proceedings/2015/january/distributed-lethality.

Rubel, Capt. Robert C., USN (Ret.). "Connecting the Dots: Capital Ships, the Littoral, Command of the Sea, and the World Order." *Naval War College Review* 68, no. 4 (Autumn 2015): 46–62.

———. "Count Ships Differently." U.S. Naval Institute *Proceedings* 146, no. 6 (June 2020). https://www.usni.org/magazines/proceedings/2020/june/count-ships-differently.

———. "The Future of the Future of Aircraft Carriers." *Naval War College Review* 64, no. 4 (Autumn 2011): 13–16.

Saunders, Commo. Stephen, RN. *Jane's Fighting Ships 2011–2012*, 114th ed. Alexandria, VA: Jane's Information Group, 2011.

Schulte, John C. "An Analysis of the Historical Effectiveness of Anti-Ship Cruise Missiles in Littoral Warfare." Master's thesis, Naval Postgraduate School, September 1994.

Schwarz, Commander Martin, German navy. "Future Mine Countermeasures: No Easy Solutions." *Naval War College Review* 67, no. 3 (Summer 2014): 123–41.

Sharp, Rear Adm. Grant, USN. "Formal Investigation into the Circumstances Surrounding the Attack on the USS Stark (FFG 31) on 17 May 1987." Washington, DC: Office of Information, Navy Department (1987).

Sims, Lt. Cdr. William S., USN. "The Inherent Tactical Qualities of All-Big-Gun, One Caliber Battleships of High Speed, Large Displacement and Gun-Power." U.S. Naval Institute *Proceedings* 32, no. 10 (October 1906): 1337–66.

Sims, Rear Adm. William S., USN. "Military Conservatism," U.S. Naval Institute *Proceedings* 48, no. 3 (March 1922). https://www.usni.org/magazines/proceedings/1922/march/military-conservatism.

Sinclair, Upton. *I, Candidate for Governor: And How I Got Licked*. Berkeley, University of California Press, 1934.

Sloan, Bill. *The Ultimate Battle: Okinawa 1945—The Last Epic Struggle of World War II*. New York: Simon & Schuster, 2007.

Smith, Gordon. *Battles of the Falklands War*. London: Ian Allen, 1989.

Smith, Cdr. Robert H. Jr., USN. "The Submarine's Long Shadow." U.S. Naval Institute *Proceedings* 92, no. 3 (March 1966): 30–39.

Smithberger, Mandy. "Brass Parachutes: The Problem of the Pentagon Revolving Door." *Project on Government Oversight*, November 5, 2018.

———. "Officers Advocating for More F-35s Often Had Financial Stakes." *The Project on Government Oversight*, August 19, 2019. https://www.pogo.org/analysis/2019/08/officers-advocating-for-more-f-35-often-had-financial-stakes/.

Spaeder, Maj. Leo, USMC. "Sir, Who Am I? An Open Letter to the Incoming Commandant of the Marine Corps," *War on the Rocks*, March 28, 2019. https://warontherocks.com/2019/03/sir-who-am-i-an-open-letter-to-the-incoming-commandant-of-the-marine-corps/.

Spector, Ronald H. *Eagle Against the Sun*. New York: Random House, 1985.

Stavridis, Adm. James, USN. *Destroyer Captain: Lessons of a First Command*. Annapolis: Naval Institute Press, 2008.

Stickles, Cdr. Brendan, USN. "Twilight of Manned Flight?" U.S. Naval Institute *Proceedings* 142, no. 4 (April 2016). https://www.usni.org/magazines/proceedings/2016/april/twilight-manned-flight.

Strategic Systems Programs. "Conventional Prompt Strike." Accessed January 15, 2021. https://www.ssp.navy.mil/six_lines_of_business/cps.html.

Summers, Chris. "Stealth Ships Steam Ahead." *BBC News*, June 10, 2004. http://news.bbc.co.uk/2/hi/technology/3724219.stm.

Symonds, Craig L. *The Civil War at Sea*. Santa Barbara, CA: Praeger, 2009.

Tangredi, Capt. Sam J., USN (Ret.). "Breaking the Anti-Access Wall." U.S. Naval Institute *Proceedings* 141, no. 5 (May 2015). https://www.usni.org/magazines/proceedings/2015/may/breaking-anti-access-wall.

The Permanent Representative of the Republic of Korea to the United Nations. "Investigation Result on the Sinking of Republic of Korea ship *Cheonan*." New York: United Nations Security Council, June 4, 2010.

Thompson, Roger. *Lessons Not Learned: The U.S. Navy's Status Quo Culture*. Annapolis: Naval Institute Press, 2007.

Thornton, Richard C. *The Falklands Sting*. Washington: Brassey's, 1998.

Tillman, Barrett. "Fear and Loathing in the Post-Naval Era." U.S. Naval Institute *Proceedings* 135, no. 6 (June 2009): 16–21.

Tiron, Roxana, and Tony Capaccio. "Boeing Helicopters Gain as Lawmakers Reject Army's Planned Cut." *Bloomberg Government News*, September 12, 2019. https://about.bgov.com/news/boeing-helicopters-gain-as-lawmakers-reject-armys-planned-cut/.

Tirpak, John A. "Retired Generals Press Congress to Fund More F-35s, Discourage 'Legacy' Buy." *Air Force Magazine*, May 1, 2019. https://www.airforcemag.com/retired-generals-press-congress-to-fund-more-f-35s-discourage-legacy-buy/.

Title 10 U.S.C. 8062(b), United States Navy: Composition; Functions. https://www
.law.cornell.edu/uscode/text/10/8062.

Toll, Ian W. *Six Frigates: The Epic History of the Founding of the U.S. Navy.* New
York: W. W. Norton, 2006.

Toti, Capt. Willliam J., USN (Ret.). "The Hunt for Full-Spectrum ASW." U.S. Naval
Institute *Proceedings* 140, no. 6 (June 2014). https://www.usni.org/magazines
/proceedings/2014/june/hunt-full-spectrum-asw.

Train, Admiral Harry D. II, USN (Ret.). "An Analysis of the Falkland/Malvinas
Islands Campaign." *Naval War College Review* 41, no. 1 (Winter 1988): 33–50.

Trevett, Lt. Col. Michael F., USA (Ret.). "Naval Mine Warfare: Historic, Political-
Military Lessons of an Asymmetric Weapon." U.S. Naval Institute *Proceedings*
139, no. 8 (August 2013). https://www.usni.org/magazines/proceedings/2013/august
/naval-mine-warfare-historic-political-military-lessons-asymmetric.

Truver, Scott C. "Taking Mines Seriously: Mine Warfare in China's Near Seas." *Naval
War College Review* 65, no. 2 (Spring 2012): 30–37.

———. "Wanted: U.S. Navy Mine Warfare Champion." *Naval War College Review*
68, no. 2 (Spring 2015): 116–27.

Tuohy, William. *America's Fighting Admirals: Winning the War at Sea in World War
II.* St. Paul, MN: Zenith Press, 2007.

Turner, Adm. Stansfield, USN (Ret.). "Is the U.S. Navy Being Marginalized?" *Naval
War College Review* 56, no. 3 (Summer 2003): 97–104.

U.S. Army. "Pershing II Weapon System Operator's Manual." Washington, DC:
Headquarters, Department of the Army, June 1, 1986.

U.S. Census Bureau. "World Population." Accessed December 31, 2020. https://www
.census.gov/data/tables/time-series/demo/international-programs/historical
-est-worldpop.html.

U.S. Department of Defense. "Annual Report to Congress: Military and Security
Developments Involving the People's Republic of China." Washington, DC: Office
of the Secretary of Defense, 2014.

———. "Annual Report to Congress: Military and Security Developments Involving
the People's Republic of China." Washington, DC: Office of the Secretary of
Defense, 2015.

———. "Annual Report to Congress: Military and Security Developments Involving
the People's Republic of China." Washington, DC: Office of the Secretary of
Defense, 2020.

———. "Annual Report to Congress: Military and Security Developments Involving
the People's Republic of China." Washington, DC: Office of the Secretary of
Defense, 2021.

————. "Annual Report to Congress: Military Power of the People's Republic of China." Washington, DC: Office of the Secretary of Defense, 2008.

U.S. Department of Defense Missile Defense Agency. "A System of Elements." Last modified April 16, 2015, accessed August 11, 2015. http://www.mda.mil/system /elements.html.

U.S. Navy. "21st Century U.S. Navy Mine Warfare: Ensuring Global Access and Commerce." Washington, DC: Program Executive Office, Littoral and Mine Warfare/Expeditionary Warfare Directorate, 2009.

————. "Aircraft Carriers - CVN: U.S. Navy Fact File." Last modified October 16, 2014, accessed August 12, 2015. http://www.navy.mil/navydata/fact_display .asp?cid=4200&tid=200&ct=4

————. "Aircraft Carriers - CVN: U.S. Navy Fact File." Last modified September 17, 2020, accessed January 16, 2021. https://www.navy.mil/Resources/Fact-Files /Display-FactFiles/Article/2169795/aircraft-carriers-cvn/.

————. "Attack Submarines - SSN: U.S. Navy Fact File." Last modified November 9, 2015, accessed January 2, 2016. http://www.navy.mil/navydata/fact_display .asp?cid=4100&ct=4&tid=100.

————. "Evolved SeaSparrow Missile (ESSM) (RIM 162D): U.S. Navy Fact File." Last modified November 19, 2013, accessed August 9, 2015. http://www.navy.mil /navydata/fact_display.asp?cid=2200&tid=950&ct=2.

————. "F-14 Tomcat Fighter: U.S. Navy Fact File." Last modified February 17, 2009, accessed May 24, 2015. http://www.navy.mil/navydata/fact_display .asp?cid=1100&tid=1100&ct=1.

————. "Harpoon Missile: U.S. Navy Fact File." Last modified February 20, 2009, accessed May 24, 2015. http://www.navy.mil/navydata/fact_display .asp?cid=2200&tid=200&ct=2.

————. "MK 53—Decoy Launching System (Nulka): U.S. Navy Fact File." Last modified November 15, 2013, accessed August 9, 2015. http://www.navy.mil/navydata /fact_display.asp?cid=2100&tid=587&ct=2.

————. "RIM 116 Rolling Airframe Missile (RAM): U.S. Navy Fact File." Last modified November 19, 2013, accessed August 9, 2015. http://www.navy.mil/navydata /fact_display.asp?cid=2200&tid=950&ct=2.

————. "Standard Missile: U.S. Navy Fact File." Last modified November 15, 2013, accessed August 9, 2015. http://www.navy.mil/navydata/fact_display .asp?cid=2200&tid=1200&ct=2.

————. "Tomahawk Cruise Missile: U.S. Navy Fact File," last modified August 14, 2014, accessed May 30, 2016. http://www.navy.mil/navydata/fact_display .asp?cid=2200&tid=1300&ct=2.

———. "Vertical Launch Anti-Submarine Rocket ASROC (VLA) Missile." Last modified December 6, 2013, accessed January 3, 2016. http://www.navy.mil/navydata /fact_display.asp?cid=2200&tid=1500&ct=2.

"U.S. Navy Sees Chinese HGV as Part of Wider Threat." *Aviation Week and Space Technology.* January 27, 2014. http://aviationweek.com/awin /us-navy-sees-chinese-hgv-part-wider-threat.

U.S. Navy, U.S. Marine Corps, and U.S. Coast Guard. "A Cooperative Strategy for 21st Century Seapower." Washington, DC: U.S. Navy, U.S. Marine Corps, and U.S. Coast Guard, 2007.

———. "Advantage at Sea: Prevailing with Integrated All-Domain Naval Power." Washington, DC: Office of the Deputy Chief of Naval Operations for Warfighting Development (OPNAV N7), 2020.

Ullman, Harlan. "There is Now Only One Path to 355 Ships." U.S. Naval Institute *Proceedings* 146, no. 4 (April 2020). https://www.usni.org/magazines /proceedings/2020/april/there-now-only-one-path-355-ships.

Union of Concerned Scientists. "UCS Satellite Database." Last modified August 1, 2020, accessed January 10, 2021. http://www.ucsusa.org/nuclear_weapons_and _global_security/solutions/space-weapons/ucs-satellite-database.html#. VcTxBflVhBc.

United States Government Accountability Office. "Navy Shipyards: Actions Needed to Address the Main Factors Causing Maintenance Delays for Aircraft Carriers and Submarines." Washington, DC: United States Government Accountability Office, August 2020.

Urbina, Ian. "Stowaways and Crimes Aboard a Scofflaw Ship." *New York Times,* July 17, 2015. http://www.nytimes.com/2015/07/19/world/stowaway-crime-scofflaw-ship .html.

Urick, Robert J. *Principles of Underwater Sound,* 2nd ed. New York: McGraw-Hill, 1975.

Utt, Ronald D. *Ships of Oak, Guns of Iron: The War of 1812 and the Forging of the American Navy.* New York: Regnery History, 2012.

van der Vat, Dan. *Stealth at Sea: The History of the Submarine.* New York: Houghton Mifflin, 1995.

Vandenengel, Lt. Cdr. Jeff, USN. "100,000 Tons of Inertia," U.S. Naval Institute *Proceedings* 146, no. 5 (May 2020): https://www.usni.org/magazines/proceedings/2020 /may/100000-tons-inertia.

———. "Fighting Along a Knife Edge in the Falklands." U.S. Naval Institute *Proceedings* 145, no. 12 (December 2019). https://www.usni.org/magazines/proceedings /2019/december/fighting-along-knife-edge-falklands.

———. "Too Big to Sink." U.S. Naval Institute *Proceedings* 143, no. 5 (May 2017). https://www.usni.org/magazines/proceedings/2017/may/too-big-sink.

Vick, Alan J., Richard M. Moore, Bruce R. Pirnie, and John Stillion. *Aerospace Operations Against Elusive Ground Targets.* Santa Monica, CA: RAND, 2001.

Webb, Jim. "Momentous Changes in the U.S. Marine Corps' Force Organization Deserve Debate." *Wall Street Journal*, March 25, 2022. https://www.wsj.com/articles/momentous-changes-in-the-marine-corps-deserve-debate-reduction-david-berger-general-11648217667?mod=opinion_lead_pos10.

Werner, Ben. "Pence: No Early Retirement for USS Harry S. Truman." *USNI News*, April 30, 2019. https://news.usni.org/2019/04/30/pence-no-early-retirement-for-uss-harry-s-truman.

Wheeler, Winslow. "More Than the Navy's Numbers Could be Sinking." *Time*, December 4, 2012. http://nation.time.com/2012/12/04/more-than-the-navys-numbers-could-be-sinking/.

Wilbur, Capt. Charles H., USN (Ret.). "Remember the *San Luis!*" U.S. Naval Institute *Proceedings* 122, no. 3 (March 1996): 86–88.

Wise, Harold Lee. "One Day of War." *Naval History Magazine* 27, no. 2 (March 2013): https://www.usni.org/magazines/naval-history-magazine/2013/march/one-day-war.

Woodward, Admiral Sandy, RN. *One Hundred Days: The Memoirs of the Falklands Battle Group Commander.* Annapolis: Naval Institute Press, 1992.

Ya'ari, Rear Admiral Yedidia "Didi," Israeli navy. "The Littoral Arena: A Word of Caution." *Naval War College Review* 48, no. 2 (Spring 1995): 7–21.

Yoshihara, Toshi, and James R. Holmes. *Red Star Over the Pacific: China's Rise and the Challenge to U.S. Maritime Strategy.* Annapolis: Naval Institute Press, 2010.

Index

Figures are indicated by "f" following the page number.

About the Author

Jeff Vandenengel is a naval officer with tours on three fast-attack submarines. Winner of the 2019 Admiral Willis Lent Award as the most tactically proficient submarine department head in the Pacific Fleet, he deployed to the Western Pacific three times and to the Atlantic during the Russian invasion of Ukraine.